Building Resilient Vision Systems: Addressing Distortions in Real-World Images

Shah

CONTENTS

CHAPTER 1

INTRODUCTION

Scene understanding has been a longstanding goal in various computer vision applications, including autonomous navigation, surveillance, and aerial imaging. With the exponential growth of media generation, there is an urgent need to automate this process. A few hours of video results in hundreds of thousands of images, making manual inspection impractical. While deep learning methods have brought significant improvements, they often rely on two crucial factors: an abundance of annotated data, and limited degradations in otherwise high-resolution inputs.

As such, techniques that perform on par with humans on standardized benchmarks [57, 161], often struggle when applied in real-world scenarios where environment variables are often unconstrained, and the images captured contain a cocktail of distortions. In fact, the presence of imaging artifacts can severely impact the recognition accuracy of state-of-the-art approaches as they reduce the amount of usable information in a given image [27, 34, 35, 61, 139, 169, 190].

To tackle these challenges, one could attempt to minimize the appearance of distortions (either through a pre-processing step or by introducing constraints to the capture format) or incorporate fine-tuning to address the nuances of the application scope. However, preventing or reducing the number of artifacts might not be possible for all applications, and pre-processing might result in the introduction of further distortions for the vision task at hand. In contrast, fine-tuning comes at the expense of acquiring and annotating extensive amounts of data.

The following work aims to address these pitfalls. To deal with the effects of adverse imaging conditions we will first perform a comprehensive analysis of the impact diverse image degradations have on visual recognition and the suitability of pre-processing algorithms to steer the effects these degradations might have for high-level vision tasks(Chapters 3 and 5). Then we will develop new strategies to deal with known setbacks such as adverse environmental conditions and multiple sources of degradations in a way that allows us to improve the automatic understanding of video and image data. We will also explore learning approaches to acquire valuable representations of complex scenes without the need for annotations, exploiting instead the additional sources of information present in data streams to reduce the dependence on annotated data.

1.1   Vision in Unconstrained Scenarios

1.1.1   Object Detection

The use of mobile video-capturing devices in unconstrained scenarios offers clear advantages in a variety of areas where autonomy is just beginning to be deployed. For instance, a camera installed on a platform like an unmanned aerial vehicle (UAV) could provide valuable information about a disaster zone without endangering human lives. A flock of such devices can facilitate the prompt identification of dangerous hazards as well as the location of survivors, the mapping of terrain, and much more. Despite the potential benefits of such applications, the data captured in unconstrained scenarios are often riddled with multiple types of degradations which can have adverse effects on visual understanding approaches.

Having a real-world application such as a search and rescue drone or an autonomous driving system fail in the presence of ambient perturbations such as rain, haze, or even motion blur could have grave consequences. Keeping in mind these conditions when designing a recognition method and taking steps to address them is of major importance if one wants to achieve high performance and robustness.

An intuitive approach would be to reduce the appearance of such perturbations using image restoration and enhancement techniques popular in the area of computational photography. Having a good staple of existing and new restoration and enhancement algorithms is a nice start, but are any of them useful for the image analysis tasks at hand? Even though recent restoration algorithms perform reasonably when improving the appearance of a degraded image, many of their enhancing capabilities do not translate well to recognition tasks as the training regime is often isolated from the visual recognition aspect of the pipeline.

Taking this into consideration, we propose to bridge the gap between traditional image enhancement approaches and visual recognition tasks as a way to jointly increase the abilities of enhancement techniques for both scenarios. In line with the above objective, as part of this work, we introduce a large-scale video benchmark for assessing image restoration and enhancement for visual recognition and perform a comprehensive study on how image restoration and enhancement techniques impact scene understanding.

1.1.2 Manipulation Detection

On the other hand, while image post-processing techniques can be used to reduce the negative effects of image artifacts, they can also be used to otherwise tamper with one's understanding or perception of a scene. To this end, different techniques have been developed in the field of media forensics to classify images as forged or authentic as well as to localize and identify the tampered regions [33, 42, 81, 101, 111, 113, 171, 181, 189]. However, image manipulation detection algorithms designed to identify local anomalies often rely on the manipulated regions being "sufficiently" different from the rest of the non-tampered regions in the image.

Such anomalies might not be easily identifiable in high-quality manipulations, and their use is often based on the assumption that certain image phenomena are associated with the use of specific editing tools. This makes the task of manipulation detection hard in and of itself, with state-of-the-art detectors only being able to detect a limited number of manipulation types. More importantly, in cases where the anomaly assumption does not hold, the detection of false positives in otherwise non-manipulated images becomes a serious problem.

To address this, we study the effectiveness of current state-of-the-art manipulation approaches and propose the suitability of using those same manipulation techniques to accentuate the anomalies present in manipulated regions to make them more iden-tifiable by a variety of manipulation detection methods.

1.2 Vision in Unsupervised Scenarios

It is undeniable that the introduction of learning models has been a turning point for many vision problems, over the last decade machine learning algorithms have reached human-like performance on a variety of tasks [57, 161]. However, these approaches often require humans in the loop to manually annotate large amounts of training data. While integral to the robustness and performance of deep-learning models, the process of acquiring labels can be prohibitive in most real-life scenarios.

Therefore, other methods have been proposed to address the problem of training such models without dense human-generated labels, spanning from learning with weak or sparse annotation [66, 84, 98] to completely unsupervised learning of temporal action segmentation [1, 11, 85, 147]. The work described here deals with the latter problem, the unsupervised learning of action segments from unlabeled video data, which can be framed as the task of unsupervised temporal action segmentation. This follows the idea that, given a set of videos all capturing the same activity, it should be possible to identify temporal segments with similar sub-actions across all videos.

Previous work [11, 85, 147] has shown that the temporal attributes can play a critical role in this task, for example, in videos showing how to make pancakes the process of cracking eggs should not only look visually similar across multiple scenes but would generally occur before other tasks such as mixing the eggs into the batter or pouring the batter onto the griddle. So, it can be assumed that actions, at least in task-oriented videos, not only share certain visual features but also occur in a similar temporal space.

Consequently, temporal prediction has been established as a way to achieve a deeper understanding of the data as it requires an implicit understanding of the structure of the observed visual features and the rules they follow while they change over time. Given this, recent approaches in the field usually extract the temporal properties of the data by learning a strong temporal regularization [85, 147], which helps to find clusters over time but may result in a lower ability to identify segments based on their visual representation.

To address this, we explore the possibility of joining visual and temporal features in order to exploit the motion and object structures cues that can be observed in a video sequence (and might be missed by solely evaluating the video frames independent of each other). Such feature representation space would allow us to detect relevant action clusters based on their visual as well as their temporal appearance.

CHAPTER 2

RELATED WORK

2.1 Detection in Unconstrained Scenarios

2.1.1 Unconstrained Object Classification Datasets

For many years the areas of image restoration and enhancement have been a
topic of interest in computational photography, leading to benchmark datasets that
are mainly used for the qualitative evaluation of image appearance. These include
very small test image sets such as Set5 [9] and Set14 [199], the set of blurred images introduced by Levin et al. [94], and the DIVerse 2K resolution image dataset
(DIV2K) [2] designed for super-resolution benchmarking. Datasets containing more
diverse scene content have been proposed including Urban100 [69] for enhancement
comparisons and LIVE1 [151] for image quality assessment. While not originally designed for computational photography, the Berkeley Segmentation Dataset has been
used by itself [69] and in combination with LIVE1 [184] for enhancement work. The
popularity of deep learning methods has increased demand for training and testing
data, which Su et al. provide as video content for deblurring work [157]. Importantly,
none of these datasets were designed to combine image restoration and enhancement
with recognition for a unified benchmark.

Most similar to the dataset that was created as part of this work are various large-
scale video surveillance datasets, especially those which provide a "fixed" overhead
view of urban scenes [43, 51, 149, 210]. However, these datasets are primarily meant
for other research areas (e.g., event/action understanding, video summarization, face

recognition) and are ill-suited for object recognition tasks, even if they share some common imaging artifacts that impair recognition.

With respect to data collected by aerial vehicles, the VIRAT Video Dataset [128] contains "realistic, natural and challenging (in terms of its resolution, background clutter, diversity in scenes)" imagery for event recognition, while the VisDrone2018 Dataset [209] is designed for object detection and tracking. Other datasets including aerial imagery are the UCF Aerial Action Data Set [126], UCF-ARG [127], UAV123 [120], UAVDT [38], Campus [140], and the multi-purpose dataset introduced by Yao et al. [187].

However, existing datasets in this area contain a limited number of frames and object categories. Table 2.1 provides a comparison of our UG^2 dataset to relevant aerial datasets. As with the computational photography datasets, none of these sets have protocols for image restoration and enhancement coupled with object recognition.

2.1.2   Image Enhancement

2.1.2.1   Visual Enhancement for Aesthetic Purposes

There is a wide variety of enhancement methods dealing with different kinds of artifacts, such as denoising (where the goal is the restoration of an image x from a corrupted observation $y = x + n$, where n is assumed to be noise with variance σ^2) [12, 28, 78, 92], compression artifact reduction (which focuses on removing blocking artifacts, ringing effects or other lossy compression-induced degradation) [37, 44, 64, 107], reflection removal [104, 152], and super-resolution (which attempts to estimate a high-resolution image from one or more low-resolution images) [36, 39, 46, 47, 79, 87, 164, 185, 186]. Other approaches designed to deal with atmospheric perturbations include dehazing (which attempts to recover the scene radiance J, the global atmospheric light A and the medium transmission t from a hazy image $I(x) = J(x)t(x) + A(1 - t(x)))$ [40, 56, 146, 162, 163], and rain removal

TABLE 2.1

COMPARISON OF THE UG2 DATASET TO RELATED AERIAL DATASETS

Dataset	Frames	Videos	Classes	Capture conditions
Inria-AILD [112]	360	—	2	Ortho-rectified aerial imagery
iSAID [194]	2,806	—	15	Satellite images (Earth Vision)
UAVDT [38]	80,000	100	3	Mobile airborne videos (UAV)
UAV123 [120]	112,578	123	—	Mobile airborne videos (UAV)
VisDrone [209]	179,264	263	10	Mobile airborne videos (UAV)
DOTA-v1.5 [182]	400,000	—	16	Satellite images (Earth Vision)
Campus [140]	929,499	—	6	Mobile airborne videos (UAV)
UG^2	**3,535,382**	**629**	**37**	Mobile airborne videos (UAV, and Glider), ground-based videos

techniques [8, 16, 74, 203].

Another popular enhancement technique is blur removal (where the objective is to recover a sharp version x' of a blurry image y without knowledge of the blur parameters) [19, 78, 88, 92, 93, 94, 95, 117]. Motion blur in videos can be a product of camera shake, movement of the object during image capture, or poor illumination conditions (which lead to the use of a low shutter speed to capture more light in the scene). In such situations, deblurring techniques can restore image quality by removing these blurring artifacts. As part of this work we perform an exploratory analysis on the impact of applying classic blind deconvolution [86] and deep learning-based deblurring techniques [121, 157] on natural scenes (where motion blur is a common artifact).

While there are plenty of approaches to improve the visual quality of a corrupted image (some of which we have discussed in the current section), most of these approaches are tailored to address a particular kind of visual aberration, and the presence of multiple problematic conditions in a single image might lead to the introduction of artifacts by the chosen enhancement technique. However, recent research has explored the possibility of handling multiple degradation types [53, 160, 192, 201].

2.1.3 Visual Enhancement for Recognition

Intuitively, if an image has been corrupted, then applying restoration techniques should help us identify objects which might have been otherwise distorted. Traditional image enhancement methods employ metrics such as Peak Signal to Noise Ratio (PSNR), Structural Similarity Index (SSIM), or Information Fidelity Criterion (IFC) to assess the perceptual quality of the reconstructed images. However, Sajjadi et al. [144] argue that the use of such metrics might not reflect the performance of some models, and propose the use of object recognition performance as an evaluation metric. They observed that methods that produced images of higher perceptual quality obtained higher classification performance despite obtaining low PSNR scores.

In agreement with this, Gondal et al. [172] observed the correlation of the perceptual quality of an image with its performance when processed by object recognition models. Similarly, Tahboub et al. [159] evaluate the impact of degradation caused by video compression on pedestrian detection. Li et al. [99] perform an in-depth analysis of diverse de-raining models where they compare them using a wide variety of metrics, including their impact on object detection. Other approaches have used visual recognition as a way to evaluate the performance of visual enhancement algorithms for tasks such as text deblurring [65, 183], image colorization [202], and single image super-resolution [122]. It is important to note that for these approaches the main purpose of their task-driven evaluation is to provide a complementary perspective on the performance of their visual enhancement method, rather than to improve machine learning-based object recognition.

Early attempts at unifying the learning objective of a high-level task like object recognition with low-level tasks include the work of Zeiler et al. [197, 198] on deblurring, and the work of Haris et al. [54] on super-resolution, where they propose an end-to-end super-resolution training procedure that incorporates detection loss as a training objective, obtaining superior object detection results compared to traditional super-resolution methods for a variety of conditions (including additional perturbations on the low-resolution images such as the addition of Gaussian noise).

While the above approaches employ object recognition in addition to visual enhancement, there are approaches designed to overlook the visual appearance of the image and instead make use of enhancement techniques to exclusively improve the object recognition performance. Sharma et al. [150] make use of dynamic enhancement filters in an end-to-end processing and classification pipeline that incorporates two loss functions (enhancement and classification). The approach focuses on improving the performance of challenging high-quality images. In contrast to this, Yim et al. [190] propose a classification architecture (comprised of a pre-processing module and a neural network model) to handle images degraded by noise. Li et al. [96] introduced a dehazing method that is concatenated with Faster R-CNN and jointly optimized as a unified pipeline. It outperforms traditional Faster R-CNN and other non-joint approaches. Singh et al. [154] propose the use of a dual directed capsule network with a dual high-resolution and targeted reconstruction loss to reconstruct very low-resolution images (16x16 pixels) in order to improve digit and face recognition.

Additional work has been undertaken in using visual enhancement techniques to improve high-level tasks such as face recognition [45, 60, 68, 102, 103, 123, 134, 167, 177, 179, 188, 191, 200] (through the incorporation of deblurring, super-resolution, and hallucination techniques) and person re-identification [76] algorithms for video surveillance data.

2.1.4 Manipulation Detection

While image manipulation approaches often aim to change the appearance of an image to either improve its aesthetic qualities or to convey/emphasize a specific message, it is commonly desirable for such manipulations to have little to no traces such that they can remain undetected by the human eye. Given the need to identify whether such manipulations have been applied to sensitive data, research in media forensics has received growing attention from academia, industry, and government agencies. Representative of this has been initiatives such as the Media Forensics (MediFor) program [166], launched by the Defense Advanced Research Projects Agency (DARPA) of the U.S. Department of Defense to push for the development of image and video integrity assessment, as well as the creation of media forensics datasets and challenges [124, 132]. A key piece of that effort was the evaluation of existing manipulation detection approaches and the creation of new ones responding to newly identified weaknesses.

Despite the growing number of manipulation detection approaches, there have been few studies providing a comprehensive analysis of the state of the field [21, 196], with most of the work focusing on methods designed to identify splice or copy-move forgeries. Zampoglou et al. [196] study traditional forgery detection methods, providing further analysis on their strengths and limitations, taking into account that data on the web would more often than not undergo multiple rounds of re-compression and scaling. They analyze the methods' performance when faced with data with varying levels of compression as well as web-based images. Similarly, Christlein et al. [21] analyze the performance of various forgery detection algorithms and their robustness to different post-processing scenarios such as the introduction of Gaussian noise, re-scaling, rotation, and JPEG compression. Nevertheless, both of these works limit their analysis to studying the detectors' performance when presented with images containing some degree of tampering. While this is a valuable step in a pipeline where the algorithms are presented with images that have been identified as forgeries, there is no study on their robustness when presented with pristine data, which is something we address in this paper.

There have been many methods proposed to detect and localize different types of manipulations. Early approaches were designed to localize specific types of forgeries through an analysis of low-level tampering artifacts such as JPEG compression [10, 81, 101, 189], anomalies in the Color Filter Array (CFA) patterns [33, 42, 49], noise variance inconsistencies [111, 113, 171], and illumination-based methods [13, 30]. While generally effective, most of these techniques depend strongly on scenario-dependent assumptions.

JPEG compression-based techniques work under the assumption that all the regions in a natural image should have undergone the same amount of compression, while spliced regions would be subject to a different compression regime than the background image they are spliced into. Similarly, CFA localization methods assume that since the CFA pattern and interpolation technique used to create a digital image varies from sensor to sensor, there would be a sufficient difference between the pattern in the manipulated and non-manipulated regions in an image to detect and localize doctored regions. However, this might not hold true in cases where the tampered regions have a similar CFA pattern (*e.g.,* in cases where both the spliced and background regions were created using the same type of sensor) or when the full image has been resized (as this would destroy the original CFA pattern of the non-tampered regions).

Noise analysis methods work under the premise that natural images have a somewhat uniform global noise pattern that is distorted by local manipulations. As such, the difference between the global and local noise characteristics could be used as an indicator of tampering. This falls through when post-processing techniques like filtering are applied after splicing as they would erode any of the anomalies these algorithms depend on. Illumination-based methods are dependent on uniform global light patterns as they rely on inconsistencies in light source direction or perspective mappings between the background and spliced regions. While often more reliable than the other methods, their performance is dependent on the size of the tampered regions, as smaller regions leave the global illumination patterns mostly unaltered.

Deep learning-based techniques have now become commonplace for manipulation detection, with early approaches patterned after the traditional methods to exploit artifacts left behind by the manipulation method, such as JPEG compression artifacts [17, 55, 73, 129], noise residuals [59, 97, 205], or boundary artifacts [7, 145]. Most of the current approaches focus on a particular source of tampering evidence or on a specific manipulation type like splicing [15, 25, 70], copy-move [24, 133, 176, 180], or in-painting [23, 211]. They obtain strong performance for the forgery method they were trained on, but with limited applicability on unseen forgery methods. To address this, some approaches aim at learning features related to multiple local anomaly types [181], while others treat forgery localization as a single-class anomaly detection problem [22, 26, 29].

Following this trend, in Section 5 we explore the use of image post-processing techniques to facilitate the identification of a wide variety of manipulations. As such, we treat any manipulation as an anomaly without differentiating between the specific kinds of tampering methods. Since our method's learning is done independently from each of the detection approaches tested, the features learned and enhanced by our network are general enough to capture and emphasize multiple types of pixel-level anomalies used by common detection approaches. To our knowledge, this work is the first one to introduce a pre-processing-based approach designed to improve the performance of existing manipulation detection approaches.

2.2   Unsupervised Action Understanding

Understanding the structure of complex activities in untrimmed videos is a challenging task in the area of action recognition. One problem here is that this task usually requires a large amount of hand-annotated minute- or even hour-long video data, but annotating such data is very time-consuming and can not easily be automated or scaled. Therefore, other methods have been proposed to address the problem of training such models without dense human-generated labels, spanning from learning with weak or sparse annotation [66, 84, 98] to completely unsupervised learning of temporal action segmentation [1, 11, 85, 147]. The work described here deals with the latter problem, the unsupervised learning of action segments from unlabeled video data, which can be framed as the task of unsupervised temporal action segmentation. This follows the idea that, given a set of videos all capturing the same activity, it should be possible to identify temporal segments with similar sub-actions across all videos.

Previous work [11, 85, 147] has shown that the temporal appearance can play a critical role in this task, as e.g., in videos showing how to make pancakes the process of cracking eggs should not only look visually similar across videos but would generally occur before other tasks such as mixing the eggs into the batter or pouring the batter onto the griddle. So, it can be assumed that actions, at least in task-oriented videos, not only share certain visual features but also occur in a similar temporal space. Recent approaches in the field usually incorporate this property by learning a strong temporal regularization [85, 147], which helps to find clusters over time but may result in a lower ability to identify segments based on their visual representation.

2.2.1 Unsupervised Learning of Temporal Sequences

While there has been plenty of work done in the area of action segmentation in video, a vast majority of the approaches are fully supervised methods relying on frame annotations [32, 66, 83, 137], or weakly supervised approaches that use some form of metadata [14, 98, 138]. Supervised models achieve high-quality temporal segmentations but their training is heavily dependent on vast amounts of good quality hand-annotated minute-or even hour-long video data which can become prohibitive in most real-life scenarios as this process can be time-consuming and can not easily be automated or scaled.

To overcome the dependence on labeled data, unsupervised methods have seen increased interest recently. Chen et al. [18] propose a domain adaptation approach based on self-supervised auxiliary tasks (SSTDA) that is able to obtain state-of-the-art performance with a reduced amount of labeled data (they show that even when using 65% of the labeled data their method still produces state-of-the-art results). One of the first unsupervised methods for action segmentation is the one proposed by Bojanovski et al. [11] based on a Frank-Wolfe optimization algorithm. Sener et al. [147] later proposed the modeling of the temporal structure of sub-activities using a combination of Generalized Mallows Model (GMM) sampling and the estimation of

the action length calculated using the frame distribution to estimate the action segmentation in complex action videos. Following a similar segmentation pipeline, Kukleva et al. [85] proposed instead a combination of temporal encoding generated using
a frame timestamp prediction network and Viterbi decoding for consistent frame-to-
cluster assignment. Another interesting take on the problem is further proposed by
Aakar et al. [1]. Different from the methods so far, the idea here is not to cluster the
feature space to identify actions but to detect action boundaries in an unsupervised
way by learning a predictive framework that uses the difference between observed
and predicted frame features as a means to determine event boundaries. As such the
approach focuses on a per video-based segmentation instead of the identification of
action classes across multiple videos.

2.2.2 Unsupervised and Self-supervised Learning of Visual Representations

Various approaches have been proposed for the learning of visual representations
without labels [75]. Particularly, in the case of learning video representations in
an unsupervised way, various approaches make use of temporal properties of the
data e.g., in the form of shuffling [119], or temporal prediction [31, 155]. Temporal
prediction has been established as a way to achieve a deeper understanding of the data
as it requires an implicit understanding of the structure of the observed visual features
and the rules they follow while they change over time [110, 155, 178]. However, it has
been pointed out that the use of traditional losses does not translate well for video
frame prediction. Srivastava et al. [155] observed that predictive models trained solely
with an MSE loss have a tendency of blurring regions with uncertainty.

This has given way to the introduction of promising alternatives such as the
adversarial loss present in Generative Adversarial Networks (GANs) [50] and Con-
ditional GANs [118], which have led to significant advances in the performance of
video prediction [31, 109, 116]. GAN-based methods have incorporated the use of U-

Net architectures [141] into their generator module given their good performance in image-to-image translation [72, 208] and image segmentation [5, 108, 141] tasks. Our proposed approach does not focus on learning feature representations as is the case in those methods but instead borrows other ideas from the field to improve the joint visual-temporal learning of an embedding space for unsupervised actions recognition.

CHAPTER 3

UG^2: A VIDEO BENCHMARK FOR ASSESSING THE IMPACT OF IMAGE
RESTORATION AND ENHANCEMENT ON AUTOMATIC VISUAL
RECOGNITION

Advances in image restoration and enhancement techniques have led to discussion
about how such algorithms can be applied as a pre-processing step to improve automatic visual recognition. In principle, techniques like deblurring and super-resolution
should yield improvements by de-emphasizing noise and increasing signal in an input
image. But the historically divergent goals of computational photography and visual
recognition communities have created a significant need for more work in this direction. To facilitate new research, we introduce a new benchmark dataset called UG^2
representing both ideal conditions and common aerial image artifacts.

The UG^2 dataset contains videos from challenging imaging scenarios containing mobile airborne cameras and ground-based captures. While we provide frame-
level annotations for the purpose of object classification, the annotations can be
re-purposed for other high-level tasks such as object detection and tracking [6]. The

Parts of this work have been published as a full paper in IEEE Winter Applications of Computer
Vision (WACV) with equal contributions from Dr. Sreya Banerjee. My contributions to this paper
were the collection and annotation of the two airborne datasets (UAV and Glider Collections),
the formulation of the super-class concept to relate our dataset to ImageNet, the development of
the evaluation scheme and metrics to measure the recognition performance of diverse enhancement
mechanisms (further discussed in Chapter 5), and the assessment of the effect of deblurring methods
on recognition (Chapter 5). As such, I restrict my discussions to these topics.
Reference: R. G. VidalMata, S. Banerjee, W. J. Scheirer, K. Grm, and V. Struc, "UG^2: a Video
Benchmark for Assessing the Impact of Image Restoration and Enhancement on Automatic Visual
Recognition," in 2018 IEEE Winter Applications of Computer Vision (WACV), pp. 1597-1606,
March 2018

(a) UAV Collection

(b) Glider Collection

(c) Ground Collection

Figure 3.1. Examples of video frames in the three UG^2 collections.

video files are provided to encourage further annotation for other vision tasks. The training and test datasets employed in the evaluation are composed of annotated frames from three different video collections. Figure 3.1 presents example frames from each of the collections.

3.1 Video Collections

3.1.1 UAV Collection

The explosive popularity of small UAVs for hobbyist photography has opened up a large opportunity for computer vision researchers to make use of related open-source video data uploaded to YouTube and other video sharing sites. This collection consists of videos recorded from small UAVs in both rural and urban areas. The videos in this collection are open-source content tagged with a Creative Commons license, obtained from the YouTube video-sharing site. Because of the source, they have different video resolutions (from 600×400 to 3840×2026) and frame rates (from 12 FPS to 59 FPS). This collection contains approximately 4 hours of aerial video distributed across 231 different videos.

For this collection, we observed 8 different video artifacts and other problems: glare/lens flare, poor image quality, occlusion, over/under exposure, camera shaking, and noise (present in some videos that use autopilot telemetry), motion blur, and fish-eye lens distortion. Additionally, this collection contains videos with problematic weather/scene conditions such as night/low light video, fog, cloudy conditions, and occlusion due to snowfall. We observed 31 different super-classes (including the non-ImageNet pedestrians class) across the training subset of these frames, from which we extracted $32,608$ object images. The cropped object images have a range of sizes, from 224×224 to 800×800.

3.1.2   Glider Collection

Similar in spirit to the UAV Collection, this collection consists of video captured from a mobile airborne source. The videos in this collection were recorded by licensed pilots of fixed-wing gliders in both rural and urban areas. It contains approximately 7 hours of aerial video, distributed across 120 different videos. The videos have frame rates ranging from 25 FPS to 50 FPS and different types of compression such as MTS, MP4, and MOV. Given the nature of this collection, the videos mostly present imagery taken from thousands of feet above the ground, further increasing the difficulty of object recognition tasks. Additionally, scenes of take-off and landing contain artifacts such as motion blur, camera shaking, and occlusion (which in some cases is pervasive throughout the videos, showcasing parts of the glider that partially occlude the objects of interest).

For the Glider Collection, we observed 6 different video artifacts and other problems: glare/lens flare, over/under exposure, camera shaking and noise, occlusion, motion blur, and fish-eye lens distortion. Furthermore, this collection contains videos with problematic weather/scene conditions such as fog, clouds, and occlusion due to rain. Across the annotated frames we observed 20 different classes (including the non-ImageNet class of pedestrians), from which we extracted $31,760$ object images. The cropped object images have a range of sizes, from 224×224 to 900×900.

3.1.3 Ground Collection

In order to provide some ground-truth with respect to problematic image conditions, we performed a video collection on the ground that intentionally induced several common artifacts. One of the main challenges for object recognition within aerial images is the difference in the scale of certain objects compared to those in the images used to train the recognition model. To address this, we recorded video of static objects (e.g., flower pots, buildings) at a wide range of distances (30ft, 40ft, 50ft, 60ft, 70ft, 100ft, 150ft, and 200ft).

In conjunction with the differing recording distances, we induced motion blur in images using an orbital shaker to generate horizontal movement at different rotations per minute (120rpm, 140rpm, 160rpm, and 180rpm). Parallel to this, we recorded video under different weather conditions (sun, clouds, rain, snow) that could affect object recognition, and employed a Sony Bloggie hand-held camera (with 1280×720 resolution and a frame rate of 60 FPS) and a GoPro Hero 4 (with 1920×1080 resolution and a frame rate of 30 FPS), whose fisheye lens introduced further distortion.

The Ground Collection contains approximately 40 minutes of video, distributed across 278 videos. The annotated frames contain 20 different ImageNet classes. Furthermore, an additional class of videos showcasing a 9×11 inch 9×9 checkerboard grid exhibiting all aforementioned distances and all intervals of rotation. The motivation for including this artificial class is to provide a reference with well-defined straight lines to assess the visual impact of image restoration and enhancement algorithms. The cropped object images have a range of sizes, from 224×224 to 1000×1000.

3.2 Annotation Format

Bounding boxes establishing object regions were manually annotated using the VATIC Video Annotation Tool [170]. We provide the VATIC annotation files for every annotated video in the dataset. As shown in Figure 3.2, a single frame might contain multiple annotated objects of different shapes and sizes.

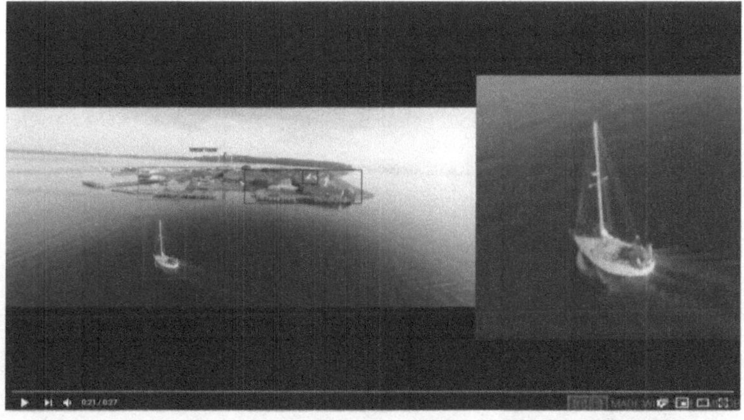

Figure 3.2. Screenshot of an UAV Collection (AV000119) annotated video. The full clip can be watched online: https://youtu.be/BA52d7X8ihw

For running classification experiments the objects were cropped from the frames in a square region of at least 224×224 pixels (a common input size for many deep learning-based recognition models), using the annotations as a guide. Each annotation in the dataset indicates the position, scale, visibility, and super-class (we define the concept of super-class in the following section) of an object in a video.

3.3 Object Classes and Their Relationship to ImageNet

A challenge presented by the objects annotated in the UAV and Glider Collections is the high variability of both object scale and rotation. These two factors make it difficult to differentiate some of the more fine-grained ImageNet categories. For example, while it may be easy to recognize a car from an aerial picture taken from hundreds (if not thousands) of feet above the ground, it might be impossible to determine whether that car is a taxi, a jeep or a sports car (see Figure 3.3). An exception to this rule is the Ground Collection where we had more control over distances from the target which made possible the fine-grained class distinction. For example, chainlink-fence and bannisters are separate classes in the Ground Collection and are not combined to form a fence super-class.

Figure 3.3. Examples of aerial images for which might not be possible to obtain a fine grained class recognition.

Thus UG^2 organizes the objects in high level classes that encompass multiple ImageNet synsets (ImageNet provides images for "synsets" or "synonym sets" of words or phrases that describe a concept in WordNet [41]) belonging to a common hierarchy (*e.g.*, the UG^2 class "car" includes the ImageNet classes "jeep", "taxi", and

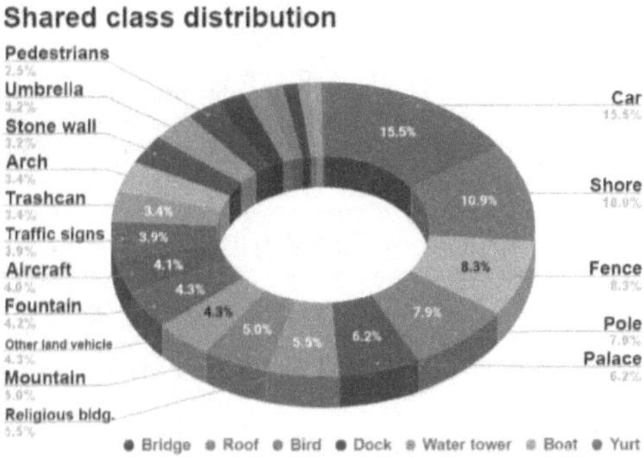

Figure 3.4. Distribution of annotated images belonging to classes shared by at least two different UG^2 collections.

"sports car"), and is used in place of such classes to account for instances in which it might be impossible to identify the fine-grained ImageNet synsets that an object belongs to. As such, each annotated image i has a single super-class label L_i which in turn is defined by a set of ImageNet synsets $L_i = \{s_1, ..., s_n\}$. Overall UG^2 is composed of 42 superclasses, which translate to over 200 ImageNet synsets.

3.4 Dataset Partitions

3.4.1 Training

The training dataset is composed of 629 videos with $3,535,382$ frames, representing 228 ImageNet [143] classes extracted from annotated frames from the three different video collections. These classes are further categorized into 37 super-classes encompassing visually similar ImageNet categories and two additional classes for pedestrian and resolution chart images. The dataset contains a subset of $152,083$ object-level annotated frames where $160,160$ objects are fully visible (out of $200,694$ total annotated frames) and the videos are tagged to indicate problematic conditions.

TABLE 3.1

SUMMARY OF THE UG^2 TRAINING DATASET

Collection	UAV	Glider	Ground	Total
Total Videos	231	120	278	629
Total Frames	1,501,675	1,840,160	193,547	3,535,382
Annotated Videos	30	30	136	196
Annotated Frames	28,263	25,246	98,574	152,083
Extracted Objects	29,826	31,760	98,574	160,160
Super-Classes	31	20	20	37

Table 3.1 summarizes the training partition of the dataset and Figure 3.4 shows the class distribution of common classes between the three collections. Over 70% of the UG^2 classes have more than 400 images and 58% of the classes are present in the imagery of at least two collections. Around 20% of the classes are present in all three collections.

3.4.2 Testing

The testing dataset is composed of 55 videos with frame-level annotations. Out of the annotated frames, $8,910$ disjoint frames were selected among the three different video collections, from which we extracted $9,187$ objects. These objects are further categorized into 42 super-classes encompassing visually similar ImageNet categories. While most of the super-classes in the testing dataset overlap with those in the training dataset, there are some classes unique to each. Table 3.2 summarizes this dataset.

TABLE 3.2

SUMMARY OF THE UG^2 TESTING DATASET

Collection	UAV	Glider	Ground	Total
Total Videos	19	15	21	55
Annotated Frames	2,814	2,909	3,187	8,910
Extracted Objects	3,000	3,000	3,187	9,187
Super-Classes [143]	28	17	20	42

CHAPTER 4

THE USE OF IMAGE MANIPULATION TECHNIQUES AS A WAY TO IMPROVE THE DETECTION OF IMAGE DOCTORING

There have been multiple techniques developed in the field of media forensics to localize and identify tampered regions in an image [33, 42, 81, 101, 111, 113, 171, 181, 189]. Such approaches often look for artifacts introduced during the forgery process as different types of manipulations tend to introduce anomalies in the image's signal. These anomalies can come in the form of inconsistent noise [25, 111, 113] and interpolation patterns [33, 42, 48, 131], blocking artifacts due to re-compression [10, 55, 71, 73, 101, 106], and other anomalies [30, 77, 80]. Learning-free methods often look for these types of anomalies by analyzing the low-level image statistics whereas more recent deep learning-based approaches try to learn intrinsic image features that could be used to differentiate manipulated and non-manipulated data.

With the rise of automated image post-processing (*e.g.*, compression, image filtering) on social media platforms that emphasize image content, the entire endeavor of image manipulation detection, as is currently practiced, is now in question. The top half of Fig. 4.1 shows an example of an input image with a splicing manipulation. While the inserted region appears to be visually coherent within the context of the image, its insertion introduced a set of anomalies that can be caught by pixel-level analysis tools. However, these tools are also sensitive to naturally occurring variations within the image, which can be incorrectly detected as tampering. This is shown in the bottom half of Fig. 4.1. Another problem exists in cases where the statistics of the spliced region are so similar to the ones extracted from the image's

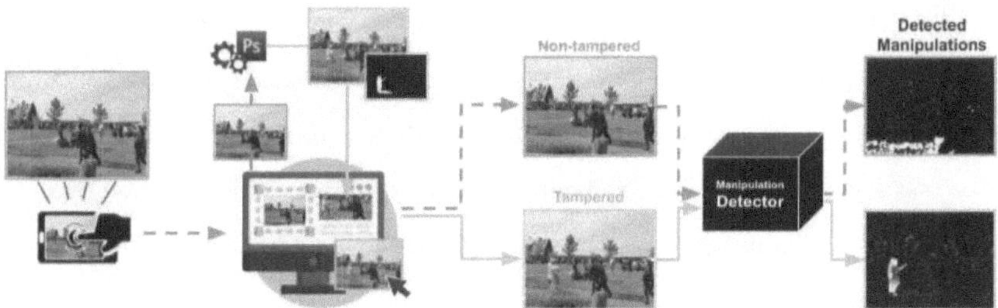

Figure 4.1. The image manipulation detection pipeline. In the right-most images presenting detection results, detected manipulated pixels are displayed in white and non-manipulated pixels in black. While manipulation detection approaches can localize the manipulated regions in a tampered image (bottom half of figure, continuous lines), the same approaches frequently misinterpret innocuous changes in non-tampered images as manipulations (top half of figure, dashed lines). The latter scenario is now widespread when analyzing images downloaded from the Internet, which almost always carry post-processing artifacts.

non-manipulated regions that the image might be able to pass undetected. As image manipulation tools get better, this is a growing concern.

One way to address these problems is to refine the detection algorithm to make it sensitive enough to catch more subtle manipulations. Another way is to better identify features that are exclusively intrinsic to tampering (if they exist). However, given the wide variety of existing (and possible future) ways to manipulate an image, generating a method that can reliably identify their effects becomes challenging. Curiously, there has been little attention paid to the performance of detection algorithms when presented with pristine images, as they are often evaluated solely on data that has varying degrees of tampering. When used for the task of distinguishing between doctored and non-doctored images in a real-life scenario, one would expect the occurrence of false positives to be relatively low. However, in our investigations into this matter, we have determined that this is not always the case.

Here we study different techniques for identifying local anomalies and review their performance on real-world forgeries as well as non-tampered data. In response to our experimental findings, we explore the use of image enhancement algorithms as a pre-processing step to improve the performance of the detection task such that anomalies in images can be more easily identified by common manipulation detection approaches.

To facilitate the work of other researchers and to allow for the replication of the experiments that appear in this section, we make a new image manipulation detection toolkit available for download, including trained models and training and evaluation source code in Python.

Our experiments and subsequent analysis are aimed at answering the following questions:

1. **How accurately can we differentiate between tampering-based anomalies and completely ordinary data?** This question relates to the situation in which the images do not contain any tampering but salient objects or naturally occurring features in the scene might be interpreted as so.

2. **Are certain local features better suited to prevent false positives?** While certain features have been proven to be a strong indicator of tampering, naturally occurring changes in the image (not related to tampering) can be mistaken for manipulations. Are certain features more prone to this?

3. **How does the preservation of pixel-level anomalies and deep-generated anomalies affect detection?** Here we seek to assess the impact of a pre-processing step to preserve/increase the presence of certain anomaly cues used by manipulation detection approaches. Is this pre-processing better for certain scenarios and worse for others?

4.1 Manipulation Detection Algorithms

4.1.1 pyIFD: A New Manipulation Detection Toolkit Comprised of Familiar Algorithms

In this section we provide a more detailed review of the learning-free local anomaly detection strategies most commonly used in practice. These approaches remain viable even in the era of deep learning, and in some cases are superior to the machine learning-based ones because they are not dependent on training data tied to specific settings. Reference implementations for these approaches that can be incorporated into a large-scale evaluation framework have been hard to come by, with many implemented in legacy Matlab or Java code [195]. To support our own study, and as a service to the media forensics community, we created the Python-based Image Forgery Detection (pyIFD) toolkit[1]. The pyIFD library is open source and written in Python to provide a cross-platform package to researchers and practitioners. It is available via PyPi to facilitate quick installation of the included algorithms. These algorithms provide strategies to estimate the anomalies left behind by different manipulation types, such as the introduction of compression artifacts, discrepancies in the image's interpolation or noise patterns, or the extraction of features that encode the differences between natural and anomalous data.

JPEG Compression Artifacts. JPEG compression takes advantage of the fact that the high-frequency information in an image is not easily perceived by the human eye. As such, it discards this type of information when encoding the image data. When performing JPEG compression the image is split into 8x8 pixel groups, each of which is further encoded separately using its own Discrete Cosine Transform (DCT). Each of these groups can be replicated by 64 cosine waves. Thus, the DCT is used to calculate the contribution of each of the waves, which often results in

---

[1]Available at `https://github.com/EldritchJS/pyIFD`.

the low-frequency ones having a larger effect than the high-frequency ones. During quantization, the high-frequency data on each of the blocks is removed by dividing each of the coefficients by their corresponding quantization value (specified by the compressor's quantization table and the JPEG quality setting) and rounding to the nearest integer. The output is then serialized to provide encoding.

This process introduces horizontal and vertical breaks into the image known as blocking artifacts. In this work, we consider three different approaches that look at different inconsistencies caused by these types of artifacts:

- **BLK [101]**: This approach is based on the observation that one of the effects of content cloning operations is the creation of a mismatch between the added and original content in the blocking artifact grid of an image. Taking this into account, BLK models the blocking artifact grid extracted out of an image's luminance component and uses the inconsistencies in the grid to identify manipulated regions. However, it is important to note that the variations in the grid might also be a product of the image content, which makes these methods sensitive to false positives due to non-JPEG features.

- **DCT [189]**: Since different JPEG-compressors use different quantization tables, the regions within a tampered image might present different types of compression artifacts if they were compressed using different tables. This approach is based on a quantization table estimation, which is used to calculate a blocking artifact measure to identify discrepancies in the local JPEG quantization history of each block.

- **ELA [81]** (Error Level Analysis): Given that JPEG is a lossy format, as an image undergoes consecutive rounds of compression, it loses more and more of its high-frequency content up to the point where further re-compressions cause no change (local minima for error). ELA works under the assumption that manipulations added to an image would have undergone fewer re-compressions than the rest of the image. As such, the difference between these "newer" regions across consecutive rounds of compression would be higher than the ones for regions that had already reached their local minima.

Discrepancies in Color Filter Array (CFA) Interpolation Patterns. Digital images are often captured using a single sensor through which the light is filtered using a CFA. As this would produce a single value per pixel, the images are then transformed into three channels using some interpolation algorithm. However, as the

CFA and the interpolation algorithm used to generate an image might not only vary from sensor to sensor but be sensitive to geometric transformations, content splicing manipulations could generate discrepancies in the CFA pattern of an image. Here we consider the following CFA-based methods:

- **CFA1 [33]**: This approach uses two features to localize tampering in an image. The first is based on estimating the CFA pattern of the source sensor. Discrepancies between the estimated and observed values are then used to detect local tampering. The second uses the variances between the noise of interpolated and natural pixels to determine the probability of the pixel values having been disrupted by tampering.

- **CFA2 [42]**: This approach also measures the presence of local interpolation artifacts between the interpolated and observed values. However, it does so by estimating the tampering probability in 2x2 blocks, which allows for a finer-grained localization of tampering.

Noise Variance Inconsistencies. Noise can be introduced into an image during its capture, compression, transmission, and other post-processing operations. While noise deteriorates the image quality, it tends to be uniformly distributed across the entire image. It has been observed that local manipulations may introduce inconsistencies in the image's global noise pattern, which makes that pattern a good candidate for observation when looking for tampered regions in an image. Several approaches have been proposed to analyze the noise patterns in an image to detect manipulations such as splicing, airbrushing, and warping. In this work we consider the following methods:

- **NOI1 [113]**: This approach analyzes the local noise level inconsistencies to detect and segment regions corrupted by local noise (the approach assumes a white Gaussian noise with a variance that can vary spatially). For this, the local noise is estimated by tiling the high pass diagonal wavelet coefficients with non-overlapping blocks (the blocks are assumed to be smaller than the size of the corrupted regions), with the standard deviation for each block used as a homogeneity condition to segment the image into sub-regions. This method is sensitive to variations in the local frequency spectrum, which makes it prone to generating false positives. Additionally, it is not able to find corrupted regions when the noise degradations are very small, which results in coarse localizations of the tampered areas.

- **NOI2** [111]: This approach is based on the observation that natural images in band-pass domains often display a statistical regularity: their kurtosis values tend to be close to a positive constant (projection kurtosis concentration). Taking this into account, NOI2 formulates a blind global noise estimation algorithm based on this property and extends it to estimate locally varying noise levels. The local noise estimation is then used to detect any possible spliced regions within the image. The estimations obtained from this approach are approximations of the local noise variances, which for tampering detection purposes are considered to be a sufficient indicator of splicing. However, this might not be the case for images with a large number of small-scale textures, which could be categorized as false positives.

- **NOI4** [171]: Since manipulations on an image often tend to leave visible traces in the image's noise map, this approach uses the noise variance inconsistencies between an image and its de-noised counterpart as a measure for tampering. For this, median filtering is performed on an image to obtain the difference between the original and median-filtered image: the median filter residual. The residual is then used as the image's forensic fingerprint.

4.1.2 Deep Learning-Based Methods

In addition to the methods included in our manipulation detection toolkit, we also consider two state-of-the-art deep learning-based manipulation detection approaches in this work. These types of detectors rely on learning features that describe possible manipulations. While some approaches aim to improve the detection of known artifacts, such as the ones addressed by the previously discussed approaches, others look for specific types of manipulations that are not explicitly defined or try to extract compact features that can be analyzed to identify manipulations at the patch level.

Mantra-Net [181] is a Manipulation Tracing Network for the localization of image forgeries. One of the key aspects of Mantra-Net is its ability to detect multiple types of manipulations. The Mantra-Net architecture is composed of two main blocks: (1) an Image Manipulation Trace Feature Extractor and (2) a Local Anomaly Detection Network.

Mantra-Net uses VGG [153] as the backbone for the feature extractor, which is trained to classify a hierarchy of image manipulations encompassing over 385 types of fine-grained manipulations. After training the feature extractor, the decision block is discarded and the extracted features are used as the input for the Local Anomaly Detection Network. This Local Anomaly Detection Network uses a module based on Long Short-Term Memory (LSTM), which is trained jointly with the pre-trained feature extractor. Given the feature map, the objective is to identify the dominant feature of the image and deem any feature that is sufficiently different from this dominant feature as anomalous.

We use the pre-trained weights provided in the Mantra-Net reference implementation for our evaluation. To account for the large amount of computation that the network performs, we crop the input images into smaller blocks (each block has at most 160K pixels) and process them individually before joining them together to generate the full manipulation mask.

RGBN [207] is a two-stream network based upon the Faster R-CNN [135] object detection network. Instead of detecting objects, this network is trained to localize tampered regions within a host image, covering diverse manipulation techniques such as splicing, copy-move, and removal. While one of the streams is focused on the RGB values of the analyzed images, the other parallel one is focused on the images' generation and capture noise (N). The underlying assumption is that the RGB stream specializes on strong-contrast manipulation artifacts (such as unnatural boundaries), while the N stream specializes on anomalies caused by disturbing the global image noise. The two streams are then combined through a bilinear pooling layer (hence the herein adopted RGBN name).

We use the implementation provided by the authors and follow their recipe to train RGBN, using the same dataset (COCO [105]) and synthetic manipulation generation processes [206].

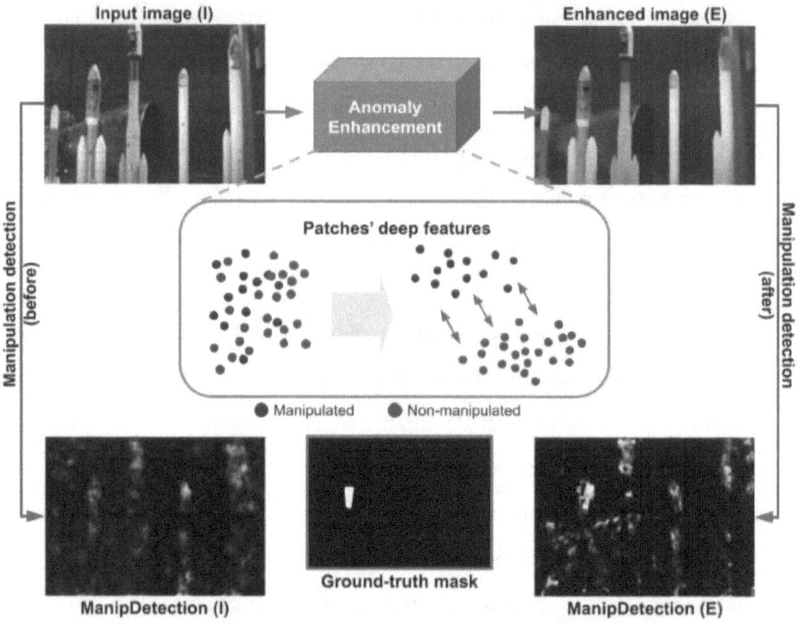

Figure 4.2. Anomaly Enhancement Pipeline. The proposed Anomaly Enhancement Network (AEN) reconstructs an input image I such that the anomalies between the tampered regions and their neighboring pixels are better emphasized in the image's enhanced version E. They can then be more easily spotted by a manipulation detection algorithm.

4.2 Anomaly Enhancement Network

With widespread access to image editing tools, the volume of doctored images present in social media has increased dramatically. In addition to this, with the increased quality of these manipulations, even humans struggle with spotting any traces left behind by tampering. One way to address these problems is to refine the detection algorithms to make them sensitive enough to catch more subtle manipulations. Another way is to identify better features exclusively intrinsic to tampering (if

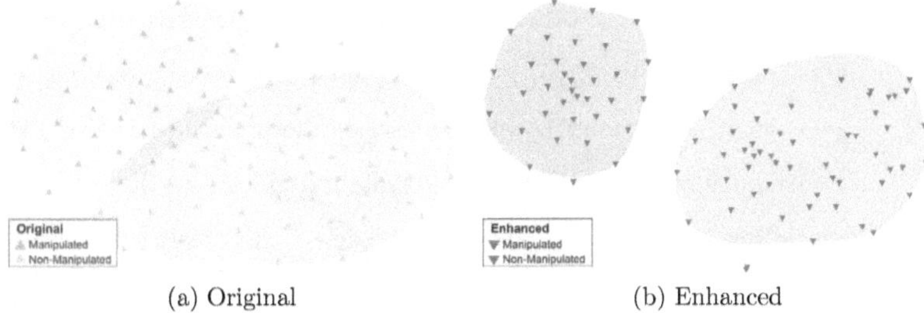

(a) Original   (b) Enhanced

Figure 4.3. Visualization using t-SNE [175] for non-linear dimensionality reduction of the embedded features of manipulated and non-manipulated patches extracted from one of the images from the MFC18 Dev Dataset. Our proposed AEN method is able to increase the distance between the manipulated and non-manipulated features as well as promote tighter clusters for patches belonging to the same class.

they exist). However, identifying these features is a challenging task given the wide variety of existing (and possible future) ways to manipulate an image. Curiously, there has been little attention paid to finding and learning to enhance image features as a pre-processing step to improve the performance of preexisting manipulation detectors.

As such, we study the suitability of one such method, and propose a novel Anomaly Enhancement Network (AEN) to be introduced as a pre-processing step in a manipulation detection pipeline (see Fig. 4.2). We base our work on the observation that in embedding spaces, image patches containing manipulated pixels often show a different distribution than the rest of the (non-manipulated) pixels in the analysed image (see Fig. 4.3a). However, the separation between the distributions of manipulated and non-manipulated data might not be as prominent depending on the type, area, and context of the manipulation.

Motivated by this, our pre-processing tool seeks to emphasize the subtle local differences between natural and tampered pixels in an embedding space to both increase the distance between their distributions and make their distributions more individually cohesive (see Fig. 4.3b). We propose to do this through a learned data-driven image content reconstruction, which preserves the original pixel-level discrepancies that manipulation detectors look for, while introducing further disparities in the tampered regions, as a way to facilitate the detection of more subtle manipulations.

Consequently, when reconstructing each region in the image, we consider three main objectives:

1. **Content preservation**: Given the different sensitivities of the manipulation detection approaches, it is important to preserve as much of the pixel-level information in the reconstructed regions in order to avoid eroding any of the anomalies that they might be tuned to find.

2. **Intra-class cohesion**: The loss of information and the possible introduction of noise and blurring artifacts during reconstruction has the potential of making non-manipulated regions similar to their tampered counterparts. To alleviate this effect we seek to increase the similarity between regions containing the same type of data (manipulated or non-manipulated) in an effort at preserving the features intrinsic to each of our classes.

3. **Anomaly enhancement:** Regardless of the content of each reconstructed region, we seek to emphasize the differences between manipulated and non-manipulated pixels within the image to facilitate the identification of tampered regions.

4.2.1 Loss Function

Consider our three main objectives. While we can optimize the content restoration using a traditional pixel-level loss (*e.g.*, L_1), in order for our network to be able to emphasize the anomalies caused by tampering while ensuring the enhancement preserves the characteristics of each region, it must also learn the differences between manipulated and non-manipulated data. For this, we present a reconstruction network with a series of triplets, each containing (1) an anchor patch a of a given class $c \in \{c_1, c_2\}$ (manipulated, or non-manipulated), (2) a positive patch p belonging to the same class as a, and (3) a negative patch n, belonging to the opposite class as

a (*i.e.*, if a belongs to the manipulated class, n would be a non-manipulated patch, and vice versa).

We then seek to produce an enhanced version of our anchor patch, $\hat{a}$, that closely resembles the input anchor a in the pixel space and other members of the anchor's class (p) in a deep embedding space. To ensure the regions in the reconstructed image highlight the tampered regions, the reconstructed patch $\hat{a}$ must also diverge from patches belonging to the contrasting class in an embedding space.

The goal of training can then be summarized by minimizing the following loss function:

$$L(a, \hat{a}, p, n) = w_0 D(a, \hat{a}) + w_1 D(f(\hat{a}), f(p)) \\ -w_2 D(f(\hat{a}), f(n)).$$

(4.1)

The loss consists of three main components. The first one corresponds to the reconstruction fidelity, measuring the distance between the reconstructed patch $\hat{a}$ and the input a in a pixel-level domain. The second is the intra-class similarity, which ensures elements from the same class are close to each other in a deep embedding domain. The last one is the anomaly enhancement factor, which seeks to increase the distance between the features extracted from manipulated and non-manipulated patches.

Within the loss, $D(x, y)$ is a distance function, $f(x)$ represents the deep features extracted from a patch x using a CNN pre-trained on ImageNet [142], and w_i corresponds to the weight for each component of the loss. Here we used the Symmetric Mean Absolute Percentage Error (SMAPE) [114] as the distance metric D to normalize the distances across each of the terms of the loss and to constrain it between 0 and 1.

While our observations on the remaining sections are based on the performance of a network trained with a balanced focus on all three objectives ($w_1 = w_2 = w_3$, as per Eq. 4.1). Here we provide an analysis of the impact of each of the three

Figure 4.4. Visualization of the effects of the different learning goals on a manipulated image.

components of our loss and evaluate the performance of our network given three weight configurations, each favoring one of the main objectives of the proposed loss.

We observed that giving a larger weight to the content restoration term of our loss produces a higher quality reconstruction (see Fig. 4.4) with a slightly lower performance than when using a balanced loss (an average of 0.2-percentage-point difference for both GWL1 and AUC). Given the premise of our approach is to increase the discrepancies between anomalous and original regions, we also tested making this element of our learning the main focus during training. Interestingly while the qualitative results show a sharper contrast between manipulated and non-manipulated regions, in some cases this learning goal increased the distance between the original and enhanced features, but not the distance between the manipulated and non-manipulated data, this was reflected by a sharp decrease in the performance on AUC (with an average decrease of 2.2 percentage points).

4.2.2   Network Architecture.

Our approach is designed to re-construct an input image I while highlighting the differences between any manipulated regions and their non-manipulated vicinities. Given the success of residual neural networks in the area of image reconstruction [4, 173, 204] we use one of such architectures as the backbone for our method. Our anomaly enhancement network is inspired by the Residual Channel Attention Network (RCAN) proposed by Zhang et al. [204] The network uses a single convolutional layer for shallow feature extraction, followed by five residual attention groups (RG) with a long connection between the shallow feature layer and the convolved output of the last residual group. Each residual group contains 10 residual channel attention blocks (RCAB) connected by short skip connections. These channel attention blocks employ a channel attention mechanism that models the inter-dependencies across the different feature channels to focus on the ones providing more informative features. The output patch is then reconstructed out of the aggregated output of the last residual group and the shallow feature layer. We take advantage of the long skip connection to preserve low-frequency features while the rest of the network focuses on the more informative high-frequency channel-wise features in the input patch. This framework is then trained using the above loss (Eq. 4.1) until convergence. The model is saved at the epoch with minimal training loss.

4.3   Experimental Setup

In this section we detail the experimental framework that will let us evaluate the effectiveness of both learning-free and deep learning-based image manipulation detectors. Importantly, we emphasize performance on known clean data that has not been manipulated in a malicious way. This is a key aspect that has been missing from previous evaluations, including the DARPA MediFor program. Because it has been missing, many prior studies overemphasize the performance of the various detectors.

4.3.1 Datasets

We use two datasets from the Media Forensics 2018 Challenge (MFC18) [52] to evaluate the detection of manipulated regions. Additionally, we analyzed our method's performance when presented with pristine images extracted from ImageNet (Imagenette [63]) and the UG^2 Dataset described in Section 3.

The MFC18 Datasets these datasets consist of images manipulated by hand by a group of expert manipulators using different media editing tools, showcasing a variety of local (*e.g.*, splicing, in-painting) and global (*e.g.*, blur, sharpening) manipulation operations. For this work, we used the *MFC18 Dev* and *MFC18 Eval* datasets, which contain manipulated images as well as their reference ground-truth masks for localization, where each bit indicates the manipulated region(s) of a distinct manipulation operation step. It is important to note that the ground-truth masks used in this work do not reflect global manipulations (*e.g.*, blur) affecting the entire image.

The proposed pre-processing step for anomaly enhancement was trained using triplets of manipulated and non-manipulated patches extracted from a subset of the MFC18 Dev Dataset. After training, we processed the images from both the MFC18 Dev and Eval Datasets (757 and 236 images respectively) and fed them to a wide variety of manipulation detection approaches to evaluate their performance (see Sec. 4.1 for a description of the detection algorithms used as well as their performance).

UG^2 Glider this is a subset of 2,778 images extracted from non-tampered frames belonging to the UG^2 Dataset [168]. The frames were extracted from 38 unique videos and showcase 15 distinctive object classes. We use this dataset as the entirety of its data does not contain any type of manipulation and was subjected to minimal rounds of compression.

Imagenette this is a subset of images extracted from the ImageNet dataset belonging to 10 distinctive object classes. We are using the validation partition (consisting of 3,925 images) for our work here. While we consider this dataset to contain data that was not changed in a malicious way, as it was collected from popular sites on the Internet, some of its images might have been subjected to global manipulations to either improve the aesthetics of the images or to add a watermark to the pictures. Nevertheless, given the nature of these manipulations and their relevance to a real-world manipulation detection task, we consider them to be "clean" data and evaluate them as such.

Triplet Extraction For Training Data

As mentioned in Section 4.2, our network is presented with a series of triplets consisting of manipulated and non-manipulated patches during training. Each triplet is formed by two patches belonging to the same class (*e.g.*, manipulated), with the third patch belonging to the other class (*e.g.*, non-manipulated). For this we extract 96x96 non-overlapping patches out of the images from the MFC18 Dev Dataset. To ensure we look at interesting points in each region we use SIFT to estimate the keypoints in the image and discard any patch with a small number of detected features. A given patch x is then labeled as manipulated if it has a manipulation ratio $r \geq 0.25$, whereas it would be labeled as non-manipulated if none of the pixels had any type of local tampering ($r = 0$). Following this, we were able to extract $15,923$ manipulated and $57,224$ non-manipulated patches out of 577 images (on average we extracted 100 patches out of each image) which we adopt as the "AEN Training Set". Given we are not able to extract useful patch triplets from all the MFC18 Dev images, the remaining 180 of them remain as unseen samples. Fig. 4.5 and 4.6 provide examples of enhancement on the training set and on unseen images, respectively. Based on these figures and on Table 4.2, we observe similar results when processing both the MFC18 Dev and Eval partitions, suggesting the proposed AEN network translate well to unseen data.

These patches are then coupled as anchor a and positive pairs (a, p), and anchor and negative pairs (a, n) to form the triplets. To make sure the training remains stable we avoid selecting triplets that are too difficult, as such for a given an anchor patch x, and a candidate patch y:

$$y = \begin{cases} p, & if \ SSIM(x, y) \geq 0.5 \\ n, & if \ SSIM(x, y) \leq 0.15 \end{cases} \tag{4.2}$$

Here we use the Structural Similarity Index Measure (SSIM) [174] to measure the similarity between each patch pair since it quantifies the differences on the structural information between two images rather than solely focusing on their pixel values. Once we have extracted valid positive and negative pairs, the patch pairs are then used to form $122,709$ triplets, with $71,550$ triplets having a manipulated patch as anchor, and $51,159$ having a non-manipulated patch as anchor.

4.3.2   Evaluation Metrics

We evaluate the performance of each of the discussed algorithms following the metrics defined by the MFC18 Challenge [124]. All of the following metrics were computed through the Mediscore tool [125], which was provided by the authors of the dataset.

Most metrics depend on having binarized values for the predicted manipulation masks, whereas the localization algorithms more often than not generate a gray-scale localization mask, where the value of each pixel ranges from 0 to 255. While it is possible to binarize the predicted masks by either specifying a binarization threshold θ, the selection of such a threshold then impacts the performance and sensitivity of such metrics. Taking this into account, we use two metrics that do not depend on binary masks: Area Under the Curve (AUC) and Grayscale Weighted L1 Loss (GWL1). In summary, the higher the AUC, the better the assessed manipulation detection solution is. Contrary to the AUC, the lower the GWL1 value, the better the solution.

Area Under the Curve Given the possibility of having multiple thresholds, we can use the area under the Receiver Operating Characteristic (ROC) curve as a measure of the system's performance across a wide range of threshold values. This metric is able to indicate how much the model is capable of distinguishing between two classes. The higher the AUC the better the model is at distinguishing between each class.

Therefore, for a given ground-truth mask m, and a predicted mask $\hat{m}$, we calculate the relation between their true positive rate (TPR) and false positive rate (FPR) after binarizing the predicted mask using a range of N thresholds $\theta_n \in \{\theta_0, ..., \theta_{N-1}\}$:

$$AUC(m, \hat{m}) = \quad \frac{1}{2} \sum_{\theta_n} \left[TPR(m, \hat{m}_{\theta_{n+1}}) + TPR(m, \hat{m}_{\theta_n}) \right] \\ \times \left[FPR(m, \hat{m}_{\theta_{n+1}}) - FPR(m, \hat{m}_{\theta_n}) \right]. \tag{4.3}$$

Gray-scale Weighted L1 Loss Each of the evaluated methods outputs a gray-scale manipulation mask, where the value for each pixel is correspondent to the confidence of that pixel being manipulated. As such, we also report the Weighted L1 Loss for the gray-scale (GWL1) masks. Contrary to the AUC, the lower the GWL1 value, the better the solution.

To make the evaluation of the masks more robust to the complexity of their borders, the Mediscore tool defines a no-score region around each ground-truth mask m. This region is defined as:

$$NoScore = Dilation(m) - Erosion(m). \tag{4.4}$$

Any pixels within the no-score region are then ignored for scoring purposes.

The GWL1 for a predicted mask $\hat{m}$ in the face of a ground-truth mask m is then defined as:

$$GWL1(m, \hat{m}) = \frac{1}{ER} \sum_{i=1}^{P} w_i \frac{|m_i - \hat{m}_i|}{255}, \qquad (4.5)$$

where m_i is the i-th pixel in the mask m, $P = size(m) = size(\hat{m})$ is the number of pixels within each mask, and $ER = size(MR) + size(NotMR)$ is the number of pixels of the Evaluated Regions MR and $NotMR$. In turn, $MR = Erosion(m)$ is the region scored as the correct manipulated region, and $NotMR = m - Dilation(m)$ is the region scored as the correct non-manipulated region. Finally, the weight w_i used for each pixel within the evaluated mask is:

$$w_i = \begin{cases} 0, & \text{if } i \in Dilation(m) \text{ and } i \notin Erosion(m) \\ 1, & \text{otherwise} \end{cases} . \qquad (4.6)$$

Since the Mediscore tool does not support the analysis of non-manipulated images, when evaluating the performance of non-manipulated data (Imagenette and UG^2 datasets), we mark a small arbitrary region (10x10 pixels) in each image as manipulated and use it as ground truth. This allows us to still use Mediscore to compare and rank the different solutions when they are presented with non-manipulated images, at the cost of a small penalty for the 10x10 regions, equally applied to all the images and methods. Given the small area of this fictional tampering, any outputs passing as true positives have little impact on the final loss score.

4.4   Results

In this section, we provide our analysis for manipulation detection performance, non-manipulated data performance, and the use of anomaly enhancement as a pre-processing step.

4.4.1 Manipulation Detection Performance

As discussed in Sec. 4.1, the features used by the evaluated algorithms can be roughly divided into four categories: (i) Compression Artifacts (BLK [101], DCT [189], ELA [81]), (ii) Color Filter Array Discrepancies (CFA1 [33], CFA2 [42]), (iii) Noise Variance (NOI1 [113], NOI2 [111], NOI4 [171]), and (iv) Deep Learning Features (MantraNet [181] and RGBN [207]).

We can see in Table 4.1 the GWL1 and AUC for each of the tampering localization methods belonging to these categories, for both the MFC18 Dev and MFC18 Eval datasets (in the first and second column groups, respectively).

When looking at the GWL1 metric, noise variance-based approaches outperform other techniques. Out of all the algorithms evaluated, NOI2 consistently produces a low GWL1 score in the predicted detection masks for both datasets. Notably, the noise variance-based approaches tended to generate sparse manipulation masks when compared to the other algorithms, as can be seen in Fig. 4.7, 4.5, and 4.6.

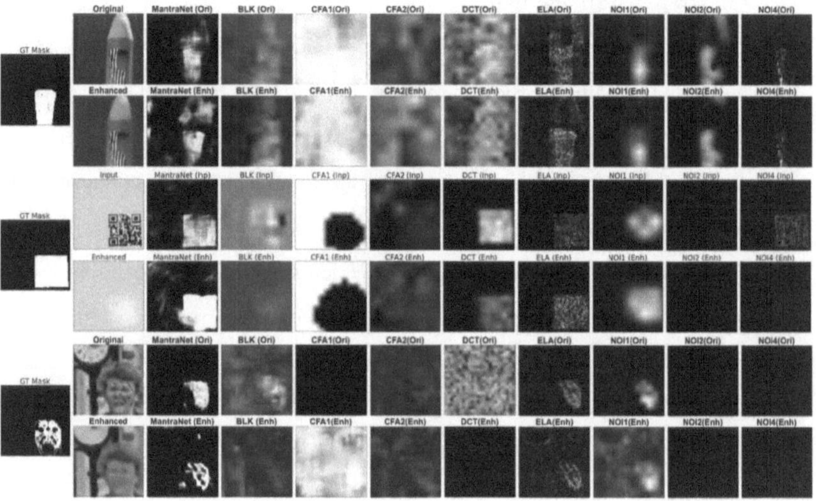

Figure 4.5. Manipulation Detection Results on the AEN Training Set. Nine different algorithms are reported over original and enhanced images from the training dataset, which contains 577 samples selected from the MFC18 Dev dataset.

TABLE 4.1

MANIPULATION DETECTION PERFORMANCE WITHOUT AEN. IN BOLD, THE BEST DATASET-WISE

RESULTS

Algorithm	MFC18 Dev Dataset		MFC18 Eval Dataset		Imagenette Dataset		UG^2 Glider Dataset	
	GWL1 (%)	AUC (%)	GWL1 (%)	AUC (%)	GWL1 (%)	AUC (%)	GWL1 (%)	AUC (%)
BLK [101]	31.23 ± 10.29	59.11 ± 28.57	43.42 ± 30.75	58.12 ± 25.14	31.52 ± 08.56	02.52 ± 01.22	28.87 ± 07.99	0.98 ± 00.23
DCT [189]	35.58 ± 15.94	59.20 ± 26.80	42.98 ± 36.03	54.19 ± 24.45	45.44 ± 12.13	03.84 ± 04.28	48.46 ± 07.12	01.55 ± 01.71
ELA [81]	11.95 ± 09.86	56.22 ± 19.76	41.54 ± 36.58	53.93 ± 16.60	15.43 ± 08.05	10.89 ± 10.70	09.68 ± 02.22	03.17 ± 01.69
CFA1 [33]	55.56 ± 27.77	55.54 ± 23.17	41.14 ± 33.50	54.88 ± 22.98	57.34 ± 24.07	10.20 ± 16.16	64.43 ± 25.40	08.60 ± 16.46
CFA2 [42]	28.39 ± 11.49	53.11 ± 24.88	27.05 ± 25.90	56.12 ± 24.80	33.01 ± 10.11	02.91 ± 02.22	36.40 ± 11.96	01.21 ± 00.68
NOI1 [113]	15.24 ± 11.68	56.52 ± 33.81	39.43 ± 36.02	54.87 ± 29.53	14.81 ± 08.25	04.14 ± 04.42	13.41 ± 05.99	01.71 ± 01.46
NOI2 [111]	**05.52 ± 10.50**	54.37 ± 18.34	**10.86 ± 17.09**	52.19 ± 11.59	05.41 ± 05.18	**20.18 ± 14.99**	**0.36 ± 01.25**	**43.95 ± 09.41**
NOI4 [171]	05.97 ± 10.27	49.30 ± 17.44	15.34 ± 17.72	47.94 ± 13.61	**02.46 ± 01.75**	19.94 ± 08.66	01.23 ± 01.53	26.63 ± 08.40
M-Net [181]	10.16 ± 09.18	**62.98 ± 23.53**	19.08 ± 25.96	**67.92 ± 22.40**	05.39 ± 02.29	06.94 ± 04.40	07.66 ± 02.63	03.03 ± 01.71
RGBN [207]	24.01 ± 29.35	52.67 ± 16.43	24.11 ± 25.54	57.61 ± 17.00	56.37 ± 31.41	21.82 ± 15.71	87.16 ± 22.70	06.42 ± 11.35

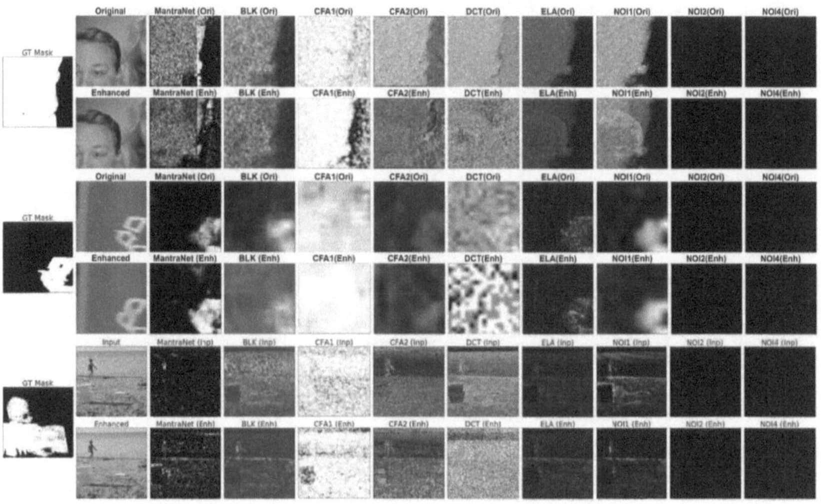

Figure 4.6. Manipulation Detection Results on Unseen Data. Nine different algorithms are reported over original and enhanced images, which were not seen by AEN during training time.

This behavior seemed to mimic the sparsity of the ground-truth manipulation masks in the MFC18 data; on average, the amount of manipulated pixels in the MFC18 Dev and MFC18 Eval datasets is relatively small, with most of the images presenting only 6% and 10% of their pixels manipulated, respectively. In comparison, the masks generated by the NOI2 algorithm for the MFC18 Dev dataset presented an average of 8.7% of their pixels detected as manipulated, and while this is close to what we observed from the ground-truth, most of the detected regions had low confidence.

On the other hand, approaches like ELA, NOI1, and Mantra-Net had a high amount of false positives in their gray-scale masks (67.7%, 95%, and 94.9% of the pixels were detected as manipulated, respectively). However, after applying different

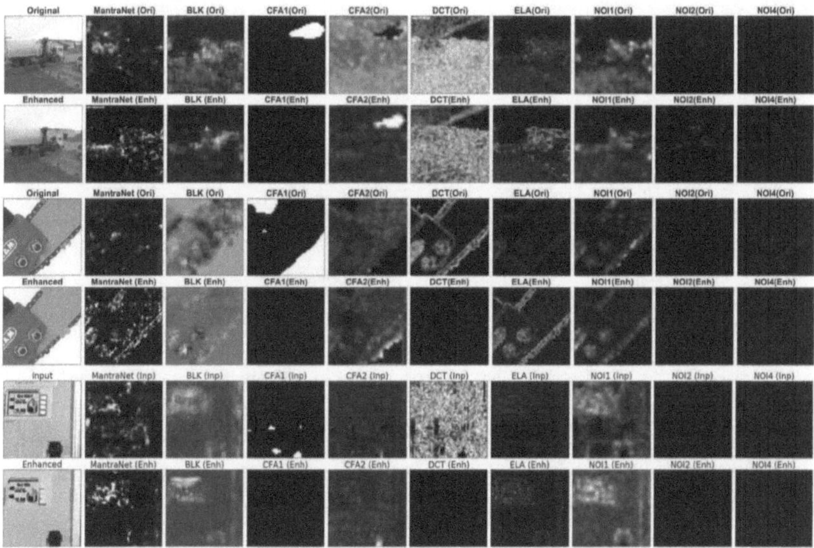

Figure 4.7. Manipulation Detection Results on the Imagenette Dataset. Nine different algorithms are reported over original and enhanced images.

confidence thresholds, their manipulation detection rates become more similar to that of the ground-truth (4.8%, 11.2%, and 3.8%, respectively), which explains their better performance in terms of AUC.

It is important to note that approaches based on CFA or DCT patterns are particularly sensitive to JPEG compression, and while they might be effective at detecting manipulations on previously uncompressed data, they often struggle with images that have undergone multiple rounds of compression (which is common in images scoured from social media) as this would erode the CFA interpolation discrepancies. This would explain their poor performance on the evaluated data, with CFA1 obtaining the worst performance on both datasets. Despite relying on similar assumptions,

CFA2 fares better than its counterpart, which could be a benefit of using a smaller localization window. Having a finer-grained localization allows it to catch any manipulation traces that were not lost after successive rounds of compression.

In turn, approaches such as ELA rely on the fact that images lose information (particularly high-frequency content) as they are re-compressed after being manipulated. Thus they target the compression residual to find tampered regions. This is especially well-suited for images captured from social media, or images that have undergone multiple rounds of manipulations such as the data present in the MFC18 datasets. Interestingly, ELA demonstrated relatively good performance (11.95% GWL1) on the MFC18 Dev dataset but struggled with the MFC18 Eval data.

4.4.2   Non-manipulated Data

Most manipulation localization approaches were designed to localize tampered regions in images that are presumed to be doctored, as such, working under the assumption that there is a manipulated region to be found within their inputs. While these approaches aim to minimize the number of false positives within each detection, as we observed from the previous section, this is not always the case, and in fact, some approaches are more prone to detect tampered regions with a low degree of confidence.

Moreover, the evaluated approaches were designed to identify features within an image that would normally be associated/emphasized by tampering. However, these discrepancies can also be naturally present in pristine data due to the natural capture process of the image, or due to the in-camera processing techniques. Taking this into account, it is important to measure the performance of manipulation detection methods not only on different types of tampered data but also on pristine samples, in order to get a full understanding of their robustness when applied to real-world scenarios. To this end, we seek to assess the behavior of the studied localization methods when presented with non-manipulated data. For this, we evaluate their performance on the Imagenette [63] and UG^2 Glider [168] datasets. We can see in Table 4.1 (last two column groups) their respective GWL1 and AUC performance.

As a subset of ImageNet [142], the Imagenette dataset was obtained by scouring the web for images representative of different object classes. As such, some of the images in this dataset might have been subjected to some kind of non-malicious manipulation, such as contrast enhancement, water-marking, or in more extreme scenarios, the removal of the object of interest's background (see the third row of Fig. 4.7, where a chainsaw has a white background). Interestingly, the performance of the manipulation detection approaches on this dataset follows similar trends to that on the UG^2 Glider Dataset, despite the latter not presenting altered content.

For both datasets, most of the approaches see a slight reduction (*i.e.*, an improvement) in their gray-scale weighted loss (GWL1) when compared to the manipulated MFC18 data. In the particular case of the best performing approach (NOI2), the GWL1 error was reduced to 0.36% over the UG^2 Glider Dataset. However, it is interesting to note that this does not translate to their performance when evaluated using AUC, for which there is a dramatic reduction, with most of the algorithms obtaining an AUC under 11%.

Similarly to what we observed in the manipulated datasets, most of these approaches detect a large number of pixels as manipulated, albeit with significantly low confidence. This impacts their AUC results, which – by definition – take into account their low confidence ranges. The positive exception relies on the robust NOI2 solution, which presents a significant reduction on its number of false positives. In the opposite direction, the more recent data-driven learning-based methods suffer from high rates of false positives, with Mantra-Net losing its previous advantage in the face of non-manipulated data.

Notably, the performance of the CFA algorithms was worse when presented with non-manipulated data, especially for the UG^2 Glider Dataset. As previously explained, these approaches rely on the presence of discrepancies between the predicted CFA interpolations. Nonetheless, their prediction of the cameras' CFA patterns might have been disrupted by the video compression present in the dataset, which led to the incongruities between the prediction and the natural pixels being erroneously detected as signs of tampering.

4.4.3 Pre-processing Anomaly Enhancement

As we saw in the previous two subsections, the performance for each of the manipulation detection algorithms varies depending on the type of features they use to determine whether a given region in an image has been tampered with or not. While their ability at detecting manipulation is dependent on the nature of the manipulation, there have been attempts to use deep learning to learn a descriptor that is able to identify multiple types of tampering [181]. Following this, we seek to assess whether the knowledge of these "global" manipulation features can be used to facilitate the detection of the anomalies used by common manipulation detection approaches. This involves the use of a neural network architecture to enhance these features before feeding the "enhanced" images to any chosen manipulation detection approach.

TABLE 4.2

MANIPULATION DETECTION PERFORMANCE ON THE MCF18 DATASETS.

Algorithm	AEN-aided GWL1 (%)		AEN-aided AUC (%)	
	MFC18 Dev	MFC18 Eval (Unseen)	MFC18 Dev	MFC18 Eval (Unseen)
BLK [101]	**28.37 ± 11.52 (↓ 02.86)**	**30.65 ± 12.45 (↓ 01.35)**	**59.33 ± 28.40 (↑ 00.22)**	**59.24 ± 27.70 (↑ 01.12)**
DCT [189]	42.65 ± 15.60 (↑ 07.08)	43.61 ± 14.40 (↑ 06.66)	49.12 ± 19.65 (↓ 10.08)	49.99 ± 16.05 (↓ 04.20)
ELA [81]	13.38 ± 09.48 (↑ 01.43)	16.66 ± 14.24 (↑ 01.45)	**57.10 ± 20.49 (↑ 00.88)**	**54.37 ± 16.11 (↑ 00.44)**
CFA1 [33]	67.76 ± 19.14 (↑ 12.20)	65.91 ± 20.51 (↑ 05.98)	**58.22 ± 21.85 (↑ 02.69)**	**58.20 ± 19.66 (↑ 03.31)**
CFA2 [42]	**27.24 ± 08.71 (↓ 01.15)**	**29.45 ± 09.91 (↓ 00.31)**	42.79 ± 22.21 (↓ 10.32)	51.17 ± 22.30 (↓ 04.95)
NOI1 [113]	16.86 ± 10.74 (↑ 01.63)	19.18 ± 13.99 (↑ 00.35)	**59.14 ± 32.32 (↑ 02.62)**	**59.27 ± 25.97 (↑ 04.39)**
NOI2 [111]	**05.37 ± 10.53 (↓ 00.14)**	**09.13 ± 16.42 (↓ 00.17)**	52.73 ± 16.64 (↓ 01.64)	51.51 ± 08.23 (↓ 00.69)
NOI4 [171]	**05.86 ± 10.26 (↓ 00.10)**	**09.68 ± 16.25 (↓ 00.16)**	52.25 ± 18.23 (↑ 02.95)	51.21 ± 12.98 (↑ 03.27)
Mantra-Net [181]	11.73 ± 08.89 (↑ 01.56)	15.35 ± 13.99 (↑ 01.82)	61.27 ± 20.45 (↓ 01.71)	60.50 ± 20.48 (↓ 07.41)
RGBN [207]	42.63 ± 26.42 (↑ 18.62)	36.49 ± 25.80 (↑ 12.38)	**57.61 ± 20.97 (↑ 04.93)**	**65.64 ± 20.07 (↑ 08.03)**

In light of this, we train the anomaly enhancement network (AEN) described in Section 4.2 and evaluate the detection performance of each of the detection algorithms after pre-processing the data (for both tampered and clean datasets) with it. As can be seen in Figs. 4.7, 4.5, and 4.6, the images produced with this new method not only preserve enough of the low-level features to attain similar manipulation masks as their original counterparts, but said manipulation masks tend to have a reduction of low-confidence detections while increasing the confidence on perceived manipulated regions.

Taking into consideration the MFC18 datasets (with manipulated images), the proposed pre-processing step was able to improve the GWL1 loss for four of the algorithms we tested (BLK, CFA2, NOI2, and NOI4), with only major deterioration in the performance for three of the remaining algorithms (DCT, CFA1, and RGBN). This can be seen through Table 4.2 (first and second data columns).

In terms of the AUC (see the last two columns of Table 4.2), the performance for the evaluated detection algorithms on the MFC18 data is ultimately low for both the baseline and enhanced images, starting at 49% and almost reaching only 68%. Regardless, our method was able to increase the AUC for six of the evaluated approaches (BLK, ELA, CFA1, NOI1, NOI4, and RGBN), with a deterioration in the performance of two of the remaining ones (DCT and CFA2).

In general, our approach seems to benefit from the short and long skip connections of the residual-based architecture, as it allows the preservation and enhancement of the noise patterns in the images, which leads to a positive impact on the noise based-methods, considering both GWL1 and AUC metrics. On the other hand, we noticed that even though we did not apply further re-compression to the enhanced images, our method struggled when dealing with JPEG compression-based methods, namely DCT and ELA, with DCT consistently being one of the algorithms most adversely impacted by our technique. This could indicate that while our approach had no issues with preserving the inconsistencies in the blocking artifact grids, it seemed to erode some of the key information that could be used by DCT and ELA to indicate distinct amounts of compression across different image regions.

As opposed to the MFC18 datasets, we do not have a ground-truth manipulation mask for the data obtained from Imagenette and the UG^2 Glider Datasets. Given that this data is presumed to be natural images, our algorithm is expected to apply a minimal set of changes to the image as there would not be any anomalies to be enhanced. In Fig. 4.7, we provide qualitative examples of how AEN worked for Imagenette.

In quantitative terms, the AEN-aided GWL1 and AUC values stayed close to the original methods' performance, as we were expecting. In the particular case of the Imagenette Dataset, whose images contain enhancements (*e.g.*, non-malicious manipulations such as contrast changes, recoloring, etc.), the GWL1 values improved for all but four methods. Regarding the more challenging Imagenette AUC values, an improvement happened for only three methods, but the AUC deterioration was mostly under one percentage point, except for two solutions (ELA and RGBN). Focusing on the UG^2 Glider Dataset (which does not contain manipulations other than the time watermarks created by the recording devices), AEN improved GWL1 for all but three methods (see the last two columns of Table 4.3).

Concerning the AUC values, it improved five methods, with a stronger contribution to the learning-based ones (Mantra-Net and RGBN). In Table 4.3, we report the impact of the proposed Anomaly Enhancement Network (AEN) as a pre-processing step over the Imagenette and UG^2 Glider Datasets, considering both the GWL1 and AUC metrics. Here, it is important to highlight that these datasets do not present maliciously manipulated content. As a consequence, we do not have ground-truth manipulation masks for the images that compose them.

While the UG^2 Glider Dataset does not contain manipulations other than the time watermarks created by the recording devices, the Imagenette Dataset does include a number of different global manipulations, such as watermarks, contrast enhancement, and recoloring. Despite the effect these global manipulations might have on the features used by the manipulation detectors, our enhancement approach was able to maintain the Imagenette GWL1 values, which stayed close to the original data's. Only four methods had an increase in GWL1 (all under 3 percentage points). Regarding the Imagenette AUC, our method improved the performance of 3 solutions. However, for the remaining manipulation detection approaches, the performance decrease was mostly under 1 percentage point, with the exception of ELA and RGBN.

TABLE 4.3

MANIPULATION DETECTION PERFORMANCE WITH AEN ON THE IMAGENETTE AND UG^2 GLIDER

DATASETS.

Algorithm	Imagenette Dataset		UG^2 Glider Dataset	
	AEN-aided GWL1 (%)	AEN-aided AUC (%)	AEN-aided GWL1 (%)	AEN-aided AUC (%)
BLK [101]	**29.02 $\pm$ 09.24 ($\downarrow$ 02.50)**	**02.52 $\pm$ 01.22 ($\uparrow$ 00.00)**	**24.45 $\pm$ 07.87 ($\downarrow$ 04.41)**	**00.98 $\pm$ 00.23 ($\uparrow$ 1.13E-05)**
DCT [189]	46.33 $\pm$ 10.48 ($\uparrow$ 00.89)	03.31 $\pm$ 03.73 ($\downarrow$ 00.53)	**44.17 $\pm$ 11.33 ($\downarrow$ 04.29)**	**03.76 $\pm$ 10.34 ($\uparrow$ 02.21)**
ELA [81]	16.70 $\pm$ 07.63 ($\uparrow$ 01.27)	05.13 $\pm$ 04.18 ($\downarrow$ 05.76)	10.82 $\pm$ 05.03 ($\uparrow$ 01.13)	02.61 $\pm$ 01.01 ($\downarrow$ 00.56)
CFA1 [33]	59.08 $\pm$ 23.69 ($\uparrow$ 01.74)	09.26 $\pm$ 15.59 ($\downarrow$ 00.94)	76.03 $\pm$ 10.49 ($\uparrow$ 11.60)	01.77 $\pm$ 06.06 ($\downarrow$ 06.83)
CFA2 [42]	**31.02 $\pm$ 09.24 ($\downarrow$ 01.99)**	02.83 $\pm$ 02.02 ($\downarrow$ 00.08)	**28.48 $\pm$ 08.56 ($\downarrow$ 07.92)**	01.05 $\pm$ 00.41 ($\downarrow$ 00.16)
NOI1 [113]	**13.65 $\pm$ 07.42 ($\downarrow$ 01.16)**	03.51 $\pm$ 03.49 ($\downarrow$ 00.63)	14.78 $\pm$ 06.53 ($\uparrow$ 01.37)	01.11 $\pm$ 00.50 ($\downarrow$ 00.61)
NOI2 [111]	**04.73 $\pm$ 05.03 ($\downarrow$ 00.68)**	**24.53 $\pm$ 14.16 ($\uparrow$ 04.35)**	**00.18 $\pm$ 00.50 ($\downarrow$ 00.18)**	**45.45 $\pm$ 06.52 ($\uparrow$ 01.50)**
NOI4 [171]	**02.24 $\pm$ 01.66 ($\downarrow$ 00.22)**	19.40 $\pm$ 08.76 ($\downarrow$ 00.54)	**00.85 $\pm$ 00.83 ($\downarrow$ 00.38)**	24.21 $\pm$ 08.44 ($\downarrow$ 02.42)
Mantra-Net [181]	08.02 $\pm$ 03.42 ($\uparrow$ 02.63)	**16.80 $\pm$ 07.52 ($\uparrow$ 09.86)**	06.40 $\pm$ 01.34 ($\downarrow$ 01.26)	**10.45 $\pm$ 05.06 ($\uparrow$ 07.42)**
RGBN [207]	**56.37 $\pm$ 31.41 ($\downarrow$ 04.02)**	21.82 $\pm$ 15.71 ($\downarrow$ 02.01)	**75.67 $\pm$ 26.22 ($\downarrow$ 11.48)**	**12.16 $\pm$ 13.11 ($\uparrow$ 05.74)**

4.5 Discussion

4.5.1 Performance on Non-manipulated Data

In a scenario where a manipulation detection approach receives as input an image with no discernible local manipulations, it would be expected for it to identify sparse (if any) regions as manipulated with a low degree of confidence. However from our experiments, we have found that this is rarely the case, and in fact, most approaches either identify large areas of the image as tampered with or in fact, point towards well-defined (salient) objects being manipulated with a high degree of confidence in the detection. This highlights the dependence that most of these algorithms have on the assumption that there should be at least two definite classes (manipulated and non-manipulated) to separate the pixels of an image, which could lead to cases where non-doctored images would be mistakenly identified as containing manipulations, with the detectors confidently pointing at arbitrary objects as the culprits. Nevertheless, by introducing a pre-processing step designed to enhance any present manipulations in an image, any change in the enhanced image serves the purpose of diminishing noise in the natural image that might otherwise be incorrectly interpreted as a sign of tampering. As such, our method was able to reduce the appearance of low-confidence pixels in the manipulation probability map, which results in cleaner predicted masks. Additionally, the intra-class cohesion property of our approach leads to stronger and more uniform manipulated segments. And while this is a property that greatly facilitates the distinction between the manipulated and non-manipulated regions in a doctored image, it can emphasize naturally occurring anomalies in an unaltered image. Despite this, our method was still able to obtain a localization performance close to the baseline for most of the approaches and even achieved an improvement of over 10 percentage points for one of the noise-based methods and for the deep-learning approaches we tested.

4.5.2 Efficacy of Different Features as Signs of Tampering

The evaluated detection techniques look into a wide variety of image features to try to identify any tampering. These approaches look at different image patterns present during the image generation process and try to identify any discrepancies in such patterns that might point to the presence of local manipulations, working under the assumption that these discrepancies are more closely related to signs of doctoring than to any naturally occurring distortions. As such, their performance at detecting manipulations of real-life manipulations is strongly tied to the type of features they use to detect tampering as well as the type of manipulation they encounter.

Despite this, we have observed that certain methods are better suited to deal with different types of manipulations across multiple datasets. Namely, we have found that approaches that analyze the noise patterns in the images and use any inconsistencies in them as a sign of tampering, consistently outperform the rest of the techniques, even the deep-learning-based ones. Since the noise patterns in the images remain stable even after multiple rounds of compression, they are well suited for analyzing images from the web that might have been saved and uploaded many times without any tampering applied to them. Moreover, these approaches also benefited from the pre-processing step, provided such a step is able to preserve the noise information in the reconstructed images.

4.5.3 Effect of Anomaly Enhancement in Detection

Following the idea that local manipulations leave fine traces that can be picked up by localization algorithms, it should be possible for an enhancement method to make such traces easier to perceive not only by the human eye but also by manipulation detection approaches. The experiments conducted in this work showcased the viability of our proposed AEN method as a pre-processing technique to improve the detection of forgeries for a variety of traditional and deep-learning-based manipulation localization algorithms.

The proposed AEN was able to improve the performance of tampering detection approaches based on the analysis of three types of anomalies: (i) JPEG compression artifacts, (ii) discrepancies in the CFA patterns, and (iii) noise variance inconsistencies. While our method was able to either improve or preserve the detection performance for most of the evaluated algorithms, noise-variance-based approaches seemed to benefit the most out of all the evaluated methods, consistently attaining high improvements on both AUC and GWL1. Despite this, our anomaly enhancement is often reflected as a blur-based distortion of the "anomalous regions" within an image, which can result in the loss of texture and other high-frequency information. While this can facilitate the recognition of anomalies to the human eyes, this can also distort valuable image statistics and throw off the estimation of the images' CFA interpolation patterns or the grid of blocking artifacts, due to the loss of high-frequency data. In some cases, this makes it impossible for the tampering detection methods to provide a correct estimate of the manipulated regions' location.

4.5.4 Anomaly Enhancement Limitations

Our method was able to identify two regions within a tampered image and reconstruct them in such a way that they are visually distinct from one another. This is reflected by one of the regions being reconstructed with a larger amount of distortions. While one would presume such a region should correspond to the manipulated area in the input image, this is not always the case. As such there are cases in which the manipulated region is reconstructed faithfully while the authentic regions in the image are distorted (see Fig. 4.8, for an example). While this effect can single out the boundary between both regions for human observers, when presented to a manipulation detection approach, it results in an inverse (and thus incorrect) prediction mask.

This might be an effect of our method's learning from disjoint image patches, which might prevent it from identifying a dominant (authentic) region within the image. We thus suggest as future research direction the incorporation of global cues into AEN as a way to address this issue.

(a) Original   (b) Enhanced

Figure 4.8. Failure scenario, where AEN preserves the spliced object and distorts the non-manipulated regions.

CHAPTER 5

THE EFFECTIVENESS OF IMAGE ENHANCEMENT TECHNIQUES IN
HIGH-LEVEL VISION TASKS

The rapid growth and ease of access to media-capturing devices offer clear advantages and challenges in various areas where autonomy is just beginning to be deployed. While a camera installed on a platform like an unmanned aerial vehicle (UAV) could provide valuable information about a disaster zone without endangering human lives, the abundance of frames captured in a single session makes the automation of their analysis a necessity. This in and of itself brings a series of challenges as a variety of environmental and capture-related artifacts often degrade such data. Moreover, the boom in the creation of media content has also increased access and interest in media editing tools, which can alter the decisions made by an automation algorithm.

As an initial step towards automation, one's first inclination might be to turn to the state-of-the-art visual recognition systems which, trained with millions of images

Sections of this chapter have been published in IEEE Winter Applications of Computer Vision (WACV) and IEEE Transactions on Pattern Analysis and Machine Intelligence (TPAMI) with equal contributions from Dr. Sreya Banerjee. My contributions to these papers were the collection and annotation of the two airborne datasets, the formulation of the super-class concept to relate our dataset to ImageNet (discussed in Chapter3), the development of the evaluation scheme and metrics to measure the recognition performance of diverse enhancement mechanisms, and the assessment of the effect of classic enhancement (blind de-convolution, and deblurring) and classification-driven methods on recognition. As such, I restrict my discussions to these topics.
References:
R. G. VidalMata, S. Banerjee, W. J. Scheirer, K. Grm, and V. Struc, "UG^2: a Video Benchmark for Assessing the Impact of Image Restoration and Enhancement on Automatic Visual Recognition," in 2018 IEEE Winter Applications of Computer Vision (WACV), pp. 1597-1606, March 2018.
R. G. VidalMata, [et. al.], "Bridging the Gap Between Computational Photography and Visual Recognition," in IEEE Transactions on Pattern Analysis and Machine Intelligence (TPAMI).

crawled from the web, would be able to identify objects, events, and human identities from a massive pool of irrelevant frames. But while these approaches have an impressive performance on popular benchmarks, it is not uncommon for them to struggle with data captured in unconstrained scenarios. One of the causes for this discrepancy in performance is that often these approaches are sensitive to the artifacts unique to the sensors used to capture the data and to the introductions of distortions product of the environment, or post-processing techniques. Even approaches designed to evaluate the presence of such perturbations (such as manipulation detection algorithms that seek the subtle traces of any post-processing-induced distortion) are constrained to specific scenarios and can be easily impacted by the introduction of multiple sources of distortions (Figure 5.3.a) making their suitability for real-world scenarios limited. Having a real-world application such as a search and rescue drone or autonomous driving system fail in the presence of ambient or post-processing perturbations could have unfortunate aftereffects.

Despite this, it remains unknown what impact many transformations have on visual recognition algorithms. To begin to answer that question, in this chapter we perform exploratory work to assess the effect common aberrations have on state-of-the-art deep learning-based recognition algorithms and analyze the impact and suitability of basic and state-of-the-art image processing algorithms used in conjunction with recognition models.

The presence of imaging artifacts has been known to have a negative impact on the performance of deep-learning-based algorithms, severely impacting the recognition accuracy of state-of-the-art approaches as they reduce the amount of usable information in a given image [27, 34, 35, 61, 139, 169, 190]. However, the extent to which different

In order to establish good baselines for classification performance, we used a selection of common deep-learning approaches to recognize annotated objects and

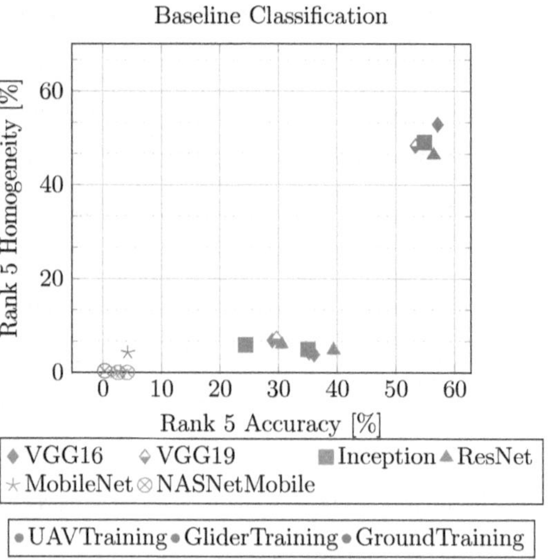

Figure 5.1. Classification rates at rank 5 for the original, unprocessed, frames for each collection in the dataset. M1 classification rate: rate of achieving at least one correctly classified synset class. M2 classification rate: rate of placing all the correct synset classes in the super-class label synset set

then considered the correct classification rate. Namely, we used the Keras [20] versions of the pre-trained networks VGG16 & VGG19 [153], Inception V3 [158], and ResNet50 [58]. We also look at two lightweight mobile networks: MobileNet [62] and NASNetMobile [212].

These experiments also serve as a demonstration of the UG^2 classification protocols. The entirety of the annotated data for all three collections is used for evaluation, with the exceptions of the pedestrian and resolution chart classes, which do not belong to any synsets recognized by the networks. For our current experiments, we restricted our analysis on UG^2 dataset to pre-trained networks. Re-training the networks with our dataset would be considered in the future.

5.1 Evaluation Metrics

The networks used for the UG^2 classification task return a list of the ImageNet synsets along with the probability of the object belonging to each of the synsets classes. However, given what we discussed in Sec. 3.3, in some cases, it is impossible to provide fine-grained labeling for the annotated objects. Consequently, most of the super-classes we defined for UG^2 are composed of more than one ImageNet synset. That is, each annotated image i has a single super-class label L_i which in turn is defined by a set of ImageNet synsets $L_i = \{s_1, ..., s_n\}$.

To measure accuracy, we observe the number of correctly identified synsets in the top-five predictions made by each pre-trained network. A prediction is considered to be correct if its synset belongs to the set of synsets in the ground-truth super-class label. We use two metrics for this. The first measures the rate of achieving at least one correctly classified synset class (M1). In other words, for a super-class label $L_i = \{s_1, ..., s_n\}$, a network is able to place at least one correctly classified synset in the top-five predictions. The second measures the rate of placing all the correct synset classes in the super-class label synset set (M2). For example, for a super-class label $L_i = \{s_1, s_2, s_3\}$, a network is able to place three correct synsets in the top-five predictions.

5.2 Baseline Classification Performance

Figure 5.1 depicts the baseline classification results for the UG^2 collections, without any pre-processing, at rank 5 (results for top 1 predictions can be found in Figure 5.2). Overall, we observed distinct differences between the results for all three collections, particularly between the airborne collections (UAV and Glider collections) and the Ground Collection. These results establish that common deep learning networks alone cannot achieve good classification rates for this dataset.

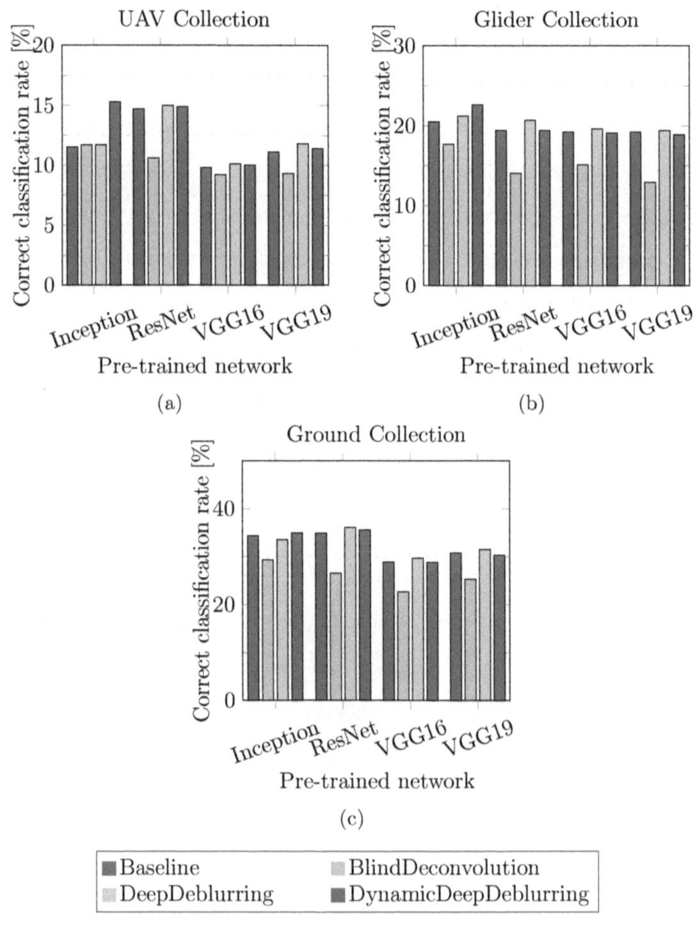

Figure 5.2. Comparison of classification rates at rank 1 after applying several deblurring techniques.

Given the very poor quality of its videos, the UAV Collection proved to be the most challenging for all networks in terms of object classification, leading to the lowest classification performance out of the three collections. While the Glider Collection shares similar problematic conditions with the UAV Collection, the images in this collection lead to a higher classification rate than those in the UAV Collection in terms of identifying at least one correctly classified synset class (metric M1). This improvement might be caused by the limited degree of movement of the gliders, since it ensures that the displacement between frames was kept more stable over time, as well as a higher recording quality (taking into consideration the camera weight limitations present in small UAVs are no longer a limiting factor for this collection's videos). The controlled Ground Collection yielded the highest classification rates, which, in an absolute sense, are still low (the highest classification rate being 65.23% for metric M1 and 52.06% for metric M2 for the testing dataset).

Additionally, we were interested in assessing the performance of lightweight classification models. As such, we employed two additional networks designed for mobile and embedded vision applications: MobileNet (v2) and NASNetMobile. The performance follows similar patterns to the ones obtained from larger classification models: (1) the UAV Collection remains the most challenging for both networks, while the Ground Collection obtained the best results out of the three collections; (2) the sharp difference in the performance on the two metrics remained. However, both mobile networks obtained poor results (when compared to the larger networks) when classifying the UG^2 data. It is likely that these models are specialized towards ImageNet-like images, and have more trouble generalizing to new data without further fine-tuning.

5.3 Enhancement Methods

The first step towards solving the problem at hand is to define a specific framework in that algorithms will operate. The challenge is to jointly optimize the two seemingly divergent but relevant tasks of (1) image enhancement and restoration and (2) object recognition. While there have been important advances in the area of computational photography [91, 121], their incorporation as a pre-processing step for higher-level tasks has received only limited attention over the past few years [150, 169]. The impact many transformations have on visual recognition algorithms remains unknown.

Ideally, for a framework to be able to jointly optimize these dual tasks, the input to the system would be a naturally corrupted image and the output is the corresponding improved image, optimized for classification performance. During training, a learning objective φ can be defined that compares the reconstructed output from the system to the one derived from baseline input. The result is then used to make further adjustments to the model if needed, in order to improve performance over the baseline. This process is shown in Figure 5.3.b.

Given the baselines established in Sec. 5.2, we now analyze the effectiveness of image enhancement approaches designed with the single goal of qualitative visual improvement as well as approaches designed for a joint/or classification-based optimization.

5.3.1 Visual Enhancement Approaches

To shed light on the effects that image restoration and enhancement algorithms have on classification, we tested classic and state-of-the-art algorithms for deblurring, including a basic blind-deconvolution approach and deep-learning-based approaches. A classic blind deconvolution algorithm can be used effectively when no information about the degradation (blur and noise) is known [86]. The algorithm restores the image and the point-spread function (PSF) simultaneously. We used Matlab's blind deconvolution algorithm, which deconvolves the image using the maximum likelihood algorithm, with a 3×3 array of 1s as the initial PSF.

With respect to state-of-the-art deep learning-based approaches, we tested the Deep Video Deblurring [157] and Deep Dynamic Scene Deblurring [121] algorithm.

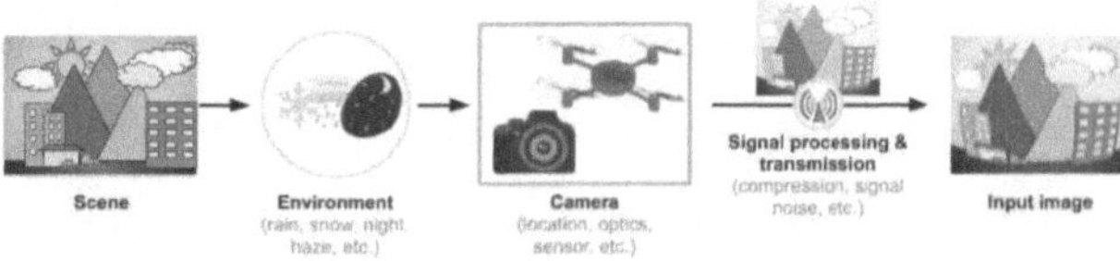

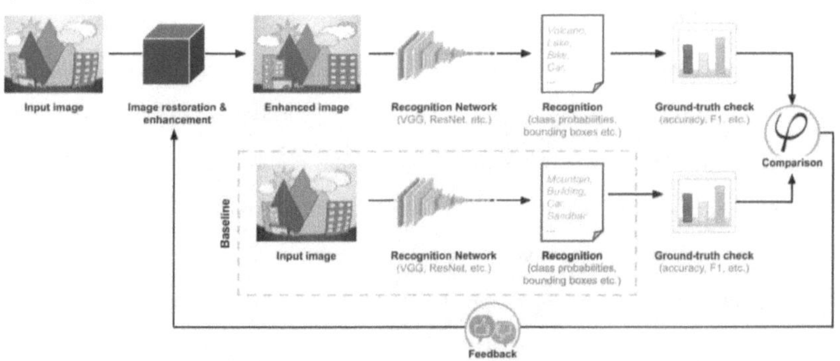

Figure 5.3. Joint Enhancement and Recognition Framework. a) Sources of image degradation during acquisition. b) and c) The proposed framework for unifying image restoration and enhancement and object recognition, supporting machine learning training and testing.

Deep Video Deblurring was designed to solely address camera shake blur. However, the results presented in [157] indicated that this method can obtain good results for other types of blur. The algorithm employs a CNN that was trained with video frames containing synthesized motion blur such that it receives a stack of neighboring frames and returns a deblurred frame. The algorithm allows for three types of frame-to-frame alignment: no alignment, optical flow alignment, and homography alignment. For our experiments, we used optical flow alignment, which was reported to have the best performance with this algorithm.

Similarly, the Deep Dynamic Scene Deblurring algorithm [121] utilizes deep learning in order to remove motion blur. Nah et al. implement a multi-scale CNN to restore blurred images in an end-to-end manner without assuming or estimating a blur kernel model. The network was trained using blurry images generated by averaging sequences (by considering gamma correction) of sharp frames in a dynamic scene with high-speed cameras. Given that this algorithm was computationally expensive, we directly applied it to the cropped object regions, rather than to the full video frame.

5.3.2 Results and Analysis

As can be observed in Figure 5.4, the behavior of the deblurring algorithms is different between the airborne and Ground collections. For the most part, the algorithms tended to improve the rate of identification for at least one correct class (M1 rate) for all of the networks for the UAV and Glider collections (Figs. 5.4b). Over 60% of the experiments reported an increase in the correct classification rate compared to that of the baseline. Conversely, for the Ground Collection, the restoration and enhancement algorithms seemed to impair the classification for all networks (Figure 5.4c), going as far as reducing the M1 identification performance by more than 16% for some experiments. More than 60% of the experiments reported a decrease in the classification rate for the Ground Collection. The property of hurting recognition performance on good quality imagery is certainly undesirable in these cases.

Further among these lines, while the M1 classification rate was increased for the airborne collections, after employing enhancement techniques the classification rate for finding all possible sub-classes in the super-class (M2 rate) was negatively impacted for all three collections. Between 53% and 68% of the experiments reported a decrease in this metric. For the UAV and Glider collections 75% and 92% of the deblurring experiments respectively had a negative impact on the classification rate for finding all possible classes.

We can also consider the performance with respect to individual networks. In these scenarios, the best overall improvement for the rate of correctly classifying at least one class in both collections was achieved by employing the Dynamic Deep Deblurring algorithm, with an improvement of 8.96% for the Inception network in the UAV case, and 6.5% for the Glider case. In contrast, Blind Deconvolution drove down performance in all of the algorithms we tested for almost all networks. For the UAV Collection, Blind Deconvolution led to a decrease of at most 6.07% in the rate of classifying at least 1 class correctly for the ResNet50 network. This behavior was also observed for the Glider and Ground collections, where it led to the highest decreases in the classification rate of both metrics for all networks. These are 7% for the ResNet50 network for the Glider Collection and 15.06% for the VGG16 network for the Ground Collection.

The results of our experiments led to some surprises. While the restoration and enhancement algorithms tended to improve the classification results for the diverse imagery included in our dataset, no approach was able to improve the results by a significant margin. Moreover, in some cases, performance degraded after image pre-processing, particularly for higher quality frames, making these kinds of pre-processing techniques unviable for heterogeneous datasets. We also noticed that different algorithms for the same type of image processing can have very different

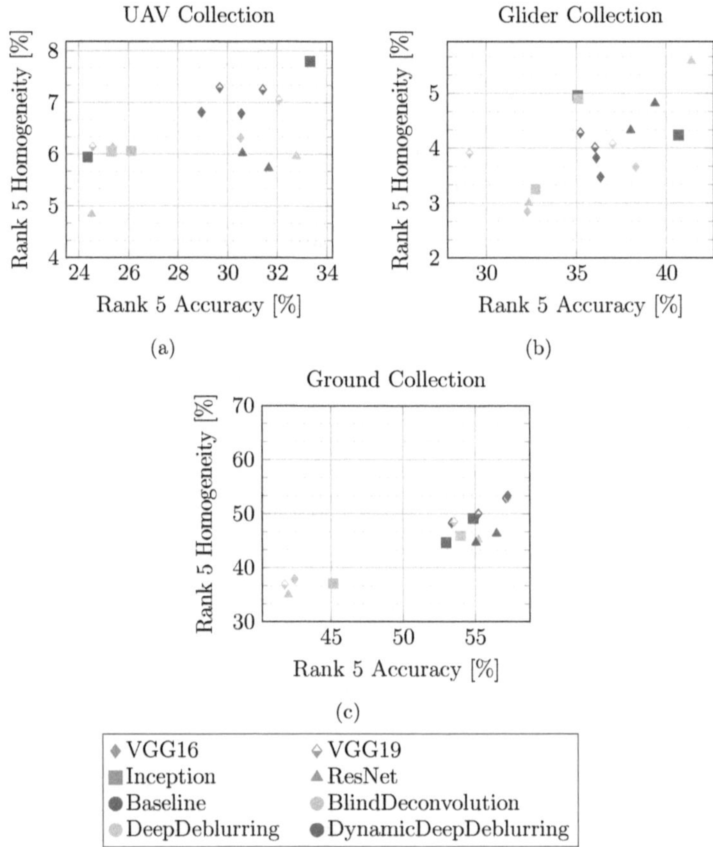

Figure 5.4. Comparison of classification rates at rank 5 for each collection after applying deblurring.

effects, as can different combinations of pre-processing and recognition algorithms. Depending on the metric considered, performance could be better or worse for various techniques. A possible reason for this can be that most of these networks were trained with images having a single type of image distortion and hence fail for images with multiple distortions from heterogeneous sources. Significant improvement can be achieved if the networks are re-trained with UG^2. However, this needs further investigation and would be done in the future.

5.4 Classification-Oriented Image Enhancement

While there are some correlations between the improvement of visual quality as perceived by humans and high-level tasks such as object classification performed by networks, research by Sharma et al. [150] suggests that image enhancement focused on improving image characteristics valuable for object recognition can lead to an increase in classification performance. Thus, we tried such a technique as an initial experiment and compared it to the baseline results. In their work, Sharma et al. develop a method to jointly optimize a convolutional neural network for enhancement and classification tasks in an end-to-end manner. This approach learns a dynamic image enhancement network with the overall goal to improve classification, but not necessarily the human perception of the image. Figure 5.5 shows the effect of the enhancement filters proposed in Sharma's work when applied to our dataset.

The proposed architecture enhances image features selectively, in such a way that the enhanced features provide a valuable improvement for a classification task. High-quality (i.e., free of visual artifacts) images are used to train a CNN to learn a configuration of enhancement filters that can be applied to an input image to yield an enhanced version that provides better classification performance.

Figure 5.5. Visual effect of the five enhancement filters proposed by Sharma et al. on the UG^2 dataset

5.4.1 Results and Analysis

Figure 5.6 shows the performance of the five enhancement filters proposed by Sharma et al. on the UG^2 dataset. It is important to note that said filters were the unaltered filters Sharma et al. trained making use of good-quality images. This is because their focus was on improving the classification performance of images with few existing perturbations. As such, the effect of improving highly corrupted images is much different from that obtained on a standard dataset of images crawled from the web. The results in Figure 5.6 establish that even existing deep learning networks designed for this task cannot achieve good classification rates for UG^2 due to the domain shift in training.

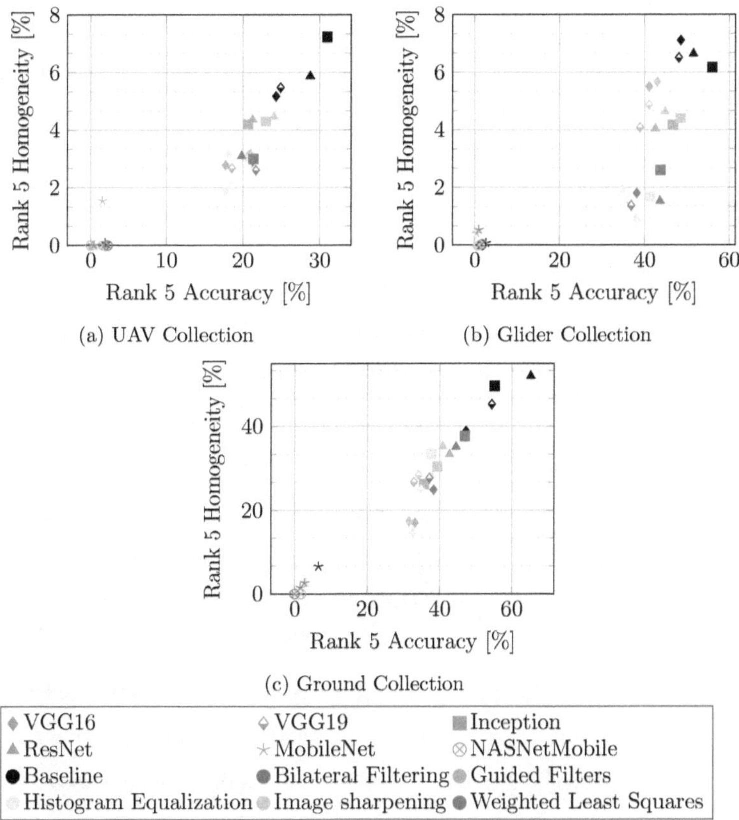

Figure 5.6. Comparison of classification rates for each collection after applying classification-driven image enhancement algorithms by Sharma et al. [150]. Markers in red indicate results on original images.

CHAPTER 6

EFFORTS TOWARDS IMPROVING ROBUSTNESS OF VISION APPROACHES
IN VIDEOS

6.1 Object Classification Improvement in Video

While interest in applying image enhancement techniques for classification purposes has started to grow, there has not been a direct application of such methods on video data. Currently, image enhancement algorithms attempt to estimate the visual aberrations a of a given image O, in order to establish an aberration-free version I of the scene captured (*i.e.*, $O = I \otimes a + n$, where n represents additional noise that might of be a byproduct of the visual aberration a). It is natural to assume that the availability of additional information — like the information present in several contiguous video frames — would enable a more accurate estimation of such aberrations, and as such, a cleaner representation of the captured scene.

For this, we adapted our evaluation method and metrics to take into account the temporal factor of the data present in the UG^2 dataset. Below we introduce the adapted training and testing datasets, as well as the evaluation metrics and baseline classification results.

6.1.1 Object Sequences in the UG^2 Dataset

To leverage both the temporal and visual features of a given scene, we divided each of the 196 videos of the original UG^2 dataset into multiple object sequences (for a total of 576 object sequences). We define an object sequence as a collection of

77

TABLE 6.1

UG^2+ OBJECT CLASSIFICATION DATASET STATISTICS.

Dataset	Collection	UAV	Glider	Ground
Training	Frames	32,608	31,760	95,096
	Object Seqs.	242	206	128
	UG^2 Classes	31	19	20
Testing	Frames	1,147	1,098	1,000
	Object Seqs.	35	32	25
	UG^2 Classes	14	7	20

multiple frames in which a given object of interest is present in the camera view. For each of these sequences, we provide frame-level annotations detailing the location of the specified object (a bounding box with its coordinates) as well as the UG^2 class.

Table 6.1 details the number of frames and object sequences extracted from each of the UG^2 collections for the training and testing datasets. An important difference between the training and testing datasets is that while some of the collections in the training set have a larger number of object sequences, that does not necessarily translate to a larger number of frames (as is the case with the UAV collection). As such, the number of frames (and thus duration) of each object sequence is not uniform across all three collections. The number of frames per object sequence can range anywhere from five frames to hundreds of frames. However, for the testing set, all of the object sequences have at least 40 frames. It is important to note that while the testing set contains imagery similar to that present in the training set, the quality of the videos might vary. This results in differences in the classification performance (more details on this are discussed in Sec. 6.1.3).

6.1.2 Evaluation Metrics

Given the nature of this task, each pre-processing algorithm is provided with a given set of object sequences rather than individual — and possibly unrelated — frames. The algorithm is then expected to make use of both temporal and visual information pertaining to each object sequence in order to provide an enhanced version of each of the sequence's individual frames. The object of interest is then cropped out of the enhanced frames and used as input to an off-the-shelf classification network. While we focus primarily on the performance of the VGG16 network (pre-trained on ImageNet), we do provide a comparative analysis of the effect of the enhancement algorithms on different classifiers. There was no fine-tuning on UG^2+ in the competition, as we were interested in making low-quality images achieve better classification results on a network trained with generally good quality data (as opposed to having to retrain a network to adapt to each possible artifact encountered in the real world). But we do look at the impact of fine-tuning in this paper.

The network provides us with a $1,000 \times n$ vector, where n corresponds to the number of frames in the object sequence, detailing the confidence score of each of the $1,000$ ImageNet classes on each of the sampled frames. To evaluate the classification accuracy of each object sequence we use Label Rank Average Precision (LRAP) [165]:

$$\text{LRAP}(y, \hat{f}) = \frac{1}{n} \sum_{i=0}^{n} \frac{1}{|y_i|} \sum_{j:y_k=1} \frac{|L_{ij}|}{rank_{ij}}$$

$$L_{ij} = \left\{ k : y_{ik} = 1, \hat{f}_{ik} > \hat{f}_{ij} \right\}$$

$$rank_{ij} = \left| \left\{ k : \hat{f}_{ik} \geq \hat{f}_{ij} \right\} \right|$$

LRAP measures the fraction of highly ranked ImageNet labels (*i.e.*, labels with the highest confidence score $\hat{f}$ assigned by a given classification network, such as VGG16) L_{ij} that belong to the true label UG^2 class y_i of a given sequence i containing n frames. A perfect score (LRAP = 1) would then mean that all of the highly ranked labels belong to the ground-truth UG^2 class. For example, if the class "shore" has two sub-classes lake-shore and sea-shore, then the top 2 predictions of the network for all of the cropped frames in the object sequence are in fact lake-shore and sea-shore. LRAP is generally used for multi-class classification tasks where a single object might belong to multiple classes. Given that our object annotations are not as fine-grained as the ImageNet classes (each of the UG^2 classes encompasses several ImageNet classes), we found this metric to be a good fit for our classification task.

6.1.3 Baseline Classification Performance

Table 6.2 shows the average LRAP of each of the collections on both the training and testing datasets without any restoration or enhancement algorithm applied to them. These scores were calculated by averaging the LRAP score of each of the object sequences y_c of a given UG^2 class C_i, for all the k classes in that particular collection D:

$$\text{AverageLRAP}(D) = \frac{1}{K} \sum_{i=0}^{K} \text{LRAP}(C_i)$$

$$\text{LRAP}(C_i) = \frac{1}{|C_i|} \sum_{c=0}^{|C_i|} \text{LRAP}(y_c, \hat{f}) \mid C_i \in D_{classes}$$

As can be observed from the training set, the average LRAP scores for each collection tend to be quite low, which is not surprising given the challenging nature of the dataset. While the Ground Collection presents a higher average LRAP, the scores from the two aerial collections are very low. This can be attributed to both aerial collections containing more severe artifacts as well as a vastly different capture

TABLE 6.2

UG² OBJECT CLASSIFICATION BASELINE STATISTICS FOR THE

IMAGENET PRE-TRAINED NETWORKS.

	UAV		Glider		Ground	
	Train	Test	Train	Test	Train	Test
VGG16	12.2%	12.7%	10.7%	33.7%	46.3%	29.4%
ResNet50	13.3%	15.1%	11.7%	28.5%	51.7%	38.7%
DenseNet201	3.9%	2.1%	3.9%	1.0%	7.5%	3.8%
MobileNetV2	1.8%	1.8%	1.5%	1.2%	6.9%	5.4%
NASNetMobile	1.2%	2.0%	1.2%	0.5%	1.0%	0.5%

viewpoint than the one in the Ground Collection (whose images would have a higher resemblance to the classification network training data). It is important to note the sharp difference in the performance of different classifiers on our dataset. While the ResNet50 [58] classifier obtained a slightly better but similar performance to the VGG16 classifier — which was the one used to evaluate the performance of the participants in the challenge — other networks such as DenseNet [67], MobileNet [62], and NASNetMobile [212] perform poorly when classifying our data. It is likely that these models are highly oriented to ImageNet-like images, and have more trouble generalizing to our data without further fine-tuning.

For the testing set, the UAV Collection maintains a low score. However, the Ground Collection's score drops significantly. This is mainly due to a higher amount of frames with problematic conditions (such as rain, snow, motion blur, or just an increased distance to the target objects), compared to the frames in the training set. A similar effect is shown on the Glider Collection, for which the majority of the videos in the testing set tended to portray either larger objects (e.g., mountains) or objects closer to the camera view (e.g., other aircraft flying close to the video-

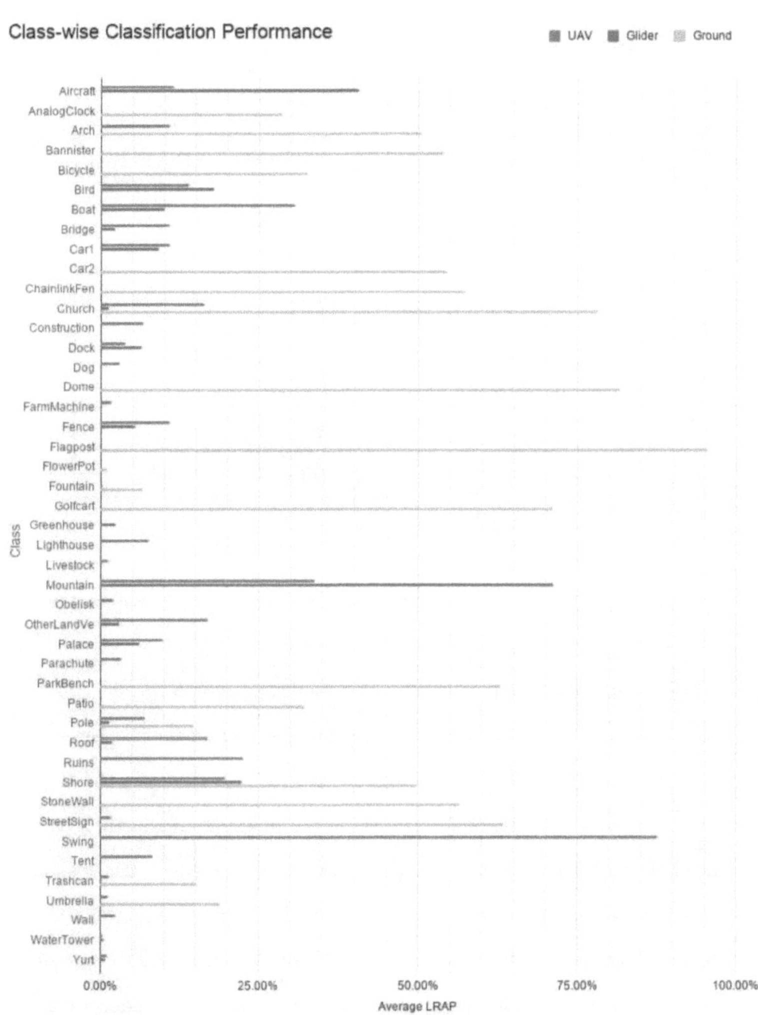

Figure 6.1. Class-wise Average LRAP performance across the three collections.

TABLE 6.3

COMPARISON OF THE VGG16 CLASSIFICATION PERFORMANCE IN

THE GLIDER COLLECTION.

Class	LRAP_train	LRAP_test
Aircraft	40.70%	72.45%
Lighthouse	—	5.05%
Mountain	71.27%	60.90%
Palace	6.18%	51.78%
Parachute	—	10.35%
Roof	1.87%	7.37%
Shore	22.30%	28.07%
TOTAL AVG	28.46%	44.12%

recording glider). Figure 6.1 shows a comparison of the classification performance of each of the UG^2 classes across all three collections. It is important to note that while some classes are particularly easy to recognize in a specific collection, the same class present in another collection might pose significant challenges. For example, the class church has a relatively high classification accuracy in the videos captured as part of the Ground Collection (in which the building area in the image is prominent even through the distortions and aberrations), whereas it has a significantly lower classification performance in the Glider Collection (where the building is captured from a considerable distance). When exclusively comparing the classification performance of the common classes of the Glider and Ground Collections, we observe a significant improvement in the average LRAP of the training set (with an average LRAP of 28.46% for the VGG16 classifier). More details on this analysis can be found in Table 6.3.

To evaluate the impact of the domain transfer between ImageNet features and our dataset, specifically looking at the disparity between the training and testing performance on the Glider collection, we fine-tuned the VGG16 network on the Glider collection training set for 200 epochs with a training/validation split of 80/20%, obtaining a training, validation, and testing accuracy of 91.67%, 27.55%, and 20.25% respectively. Once the network was able to gather more information about the dataset, the gap between validation and testing was diminished. Nevertheless, the broad difference between the training and testing scores indicates that the network has problems generalizing to the UG^2+ data, perhaps due to the large number of image aberrations present in each image.

6.2 Joint Visual-Temporal Information as a Way to Improve Unsupervised Learning of Actions in Untrimmed Sequences

Understanding the structure of complex activities in untrimmed videos is a challenging task in the area of action recognition. Research shows that humans usually understand complex activities through ongoing temporal segmentation of perceived input into meaningful segments [193]. Nevertheless, replicating this behavior in fully automated systems is a challenging problem as it requires identifying the meaningful steps in a given task and how do they logically relate to each other. Fully supervised systems have been proposed to realize such a temporal segmentation, but they rely on large amounts of training data. However, manually annotating such data is especially expensive for temporal video segmentation as this task usually requires a dense frame-based annotation. Weakly supervised approaches attempt to alleviate this by incorporating the use of additional sources of information such as speech or

Parts of this work have been published as a full paper in IEEE Winter Applications of Computer Vision (WACV). **Reference**: R. G. VidalMata, W. J. Scheirer, A. Kukleva, D. D. Cox, and H. Kuehne, "Joint Visual-Temporal Embedding for Unsupervised Learning of Actions in Untrimmed Sequences," in 2021 IEEE Winter Conference on Applications of Computer Vision (WACV).

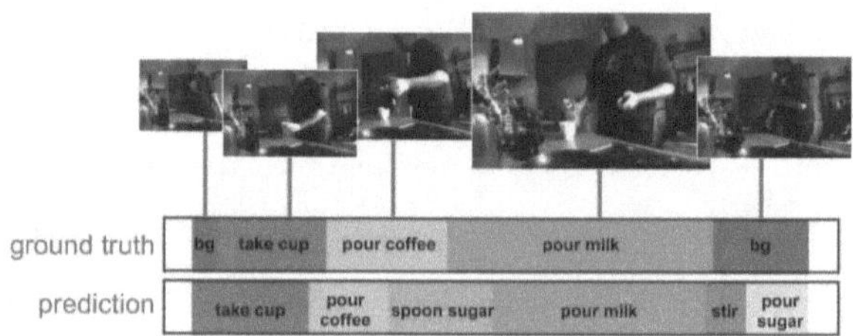

Figure 6.2. Temporal segmentation of a video from the Breakfast Actions [82] Dataset. Our approach maintains the logical ordering of sub-activities and a good estimate on their start and duration.

video captions [3, 115], but most available real-world video data —such as surveillance data— does not come with any additional modality such as audio, subtitles, or descriptive meta-data. Additionally, differences between the alignment of additional modalities like audio, subtitles, or descriptive meta-data to the video frames might prove challenging for some of these approaches [148]. This makes it difficult for such approaches to translate to real-world applications.

As was discussed in Chapter 2, other methods have been proposed to tackle the challenge of training such models without dense human-generated labels, spanning from learning with weak or sparse annotation [66, 84, 98] to completely unsupervised learning of temporal action segmentation [1, 11, 85, 147]. The work described here deals with the latter problem, the unsupervised learning of action segments from unlabeled video data, which can be framed as the task of unsupervised temporal action segmentation. This follows the idea that, given a set of videos all capturing the same activity, it should be possible to identify temporal segments with similar sub-actions across all videos.

Previous work [11, 85, 147] has shown that temporal appearance can play a critical role in this task. For instance, in videos showing how to make pancakes the process of cracking eggs should not only look visually similar across different videos but would generally occur before other tasks such as mixing the eggs into the batter or pouring the batter onto the griddle. So, it can be assumed that actions, at least in task-oriented videos, not only share certain visual features but also occur in a similar temporal space. Recent approaches usually incorporate this property by learning a strong temporal regularization [85, 147], which helps to find clusters over time but may result in a lower ability to identify segments based on their visual representation.

To address this problem, we propose a joint visual-temporal learning pipeline that combines the advantages of current temporal embedding systems with a visual embedding based on a combination of predictive visual and temporal learning tasks. To this end, we combine a state-of-the-art temporal embedding system [85], designed to estimate the relative timestamp of a given video frame, with a visual encoder-decoder pipeline that is trained on a combination of visual and temporal losses.

The intuition behind this is that the embedding should not only reconstruct the plain input signal, but should also find a reconstruction that allows for a better estimation of the respective timestamp and, thus, a better temporal reconstruction. To prevent the learning of weak visual cues, we shift the output of the visual encoder by a number of frames, which turns it into a visual prediction architecture, similar to other self-supervised models [31, 116]. Combined with the temporal loss, the model predicts a frame-representation that is optimized to give the best timestamp prediction in the temporal embedding framework. The resulting embedding space thus captures visual and temporal representations of each individual frame.

We evaluate the proposed system on three challenging standard benchmark datasets, the Breakfast dataset [82], the INRIA YouTube Instructional Videos (YTI) dataset [11], and the 50 Salads dataset [156], achieving good temporal segmentation compared to state-of-the-art approaches. Figure 6.2 shows a qualitative example that demonstrates that the proposed architecture is able to adapt well to a diverse set of action tasks.

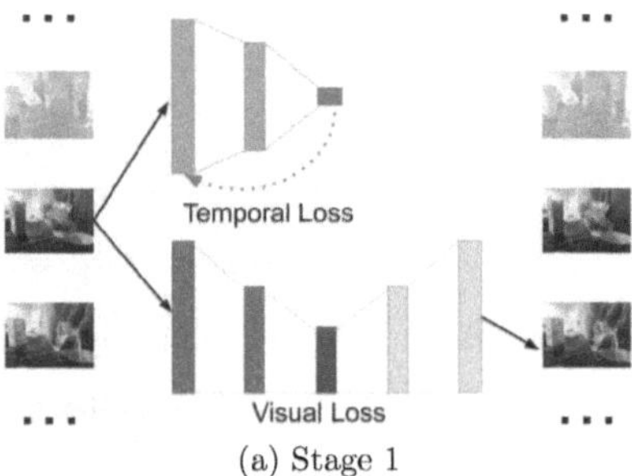

(a) Stage 1

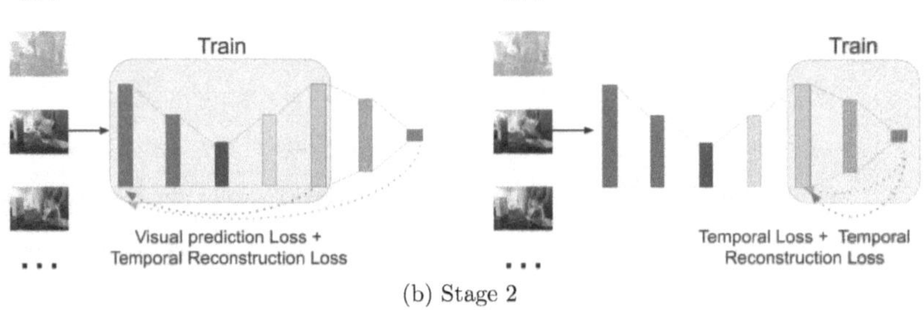

(b) Stage 2

Figure 6.3. Two-stage visual-embedding pipeline: 6.3a) Visual+Temporal embedding: next frame prediction U-Net to generate a visual-temporal embedding (output of the last down-sampling layer, denoted by a red arrow), 6.3b) Temporal discriminator: timestamp predictor MLP used to identify the loss of temporal information in the frames predicted by stage 1.

6.2.1 System Description

Given a collection of complex activity videos $D_A = \{v_i\}_{i=1}^{V}$ with V videos belonging all to the same activity class A, we want to learn k action classes C_k that appear in this activity, as well as the action class label of all the frames (N_{v_i}) of each video $v_i = \{f_n\}_{n=1}^{N_{v_i}}$. Note that the frame input used here is a pre-extracted one-dimensional feature vector, that can be derived from hand-crafted feature representations or from a network that has been trained in an unsupervised way.

To learn an embedding space for this input, we propose a two-stage pipeline (see

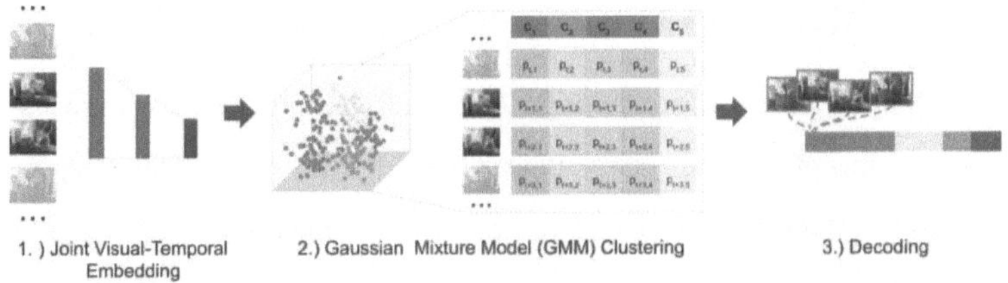

Figure 6.4. Joint visual-temporal training pipeline. If the feature embedding has a good representation of the visual and temporal attributes of each frame, the frames that cluster together would have similar temporal locations and share visual attributes that represent a given sub-activity.

Figure 6.3). The first stage (6.3a) consists of the independent training of two models: one encoding a visual embedding and the second encoding a temporal embedding only. We pre-train both models separately and then join them in the second stage (6.3b), where the temporal model becomes a discriminator for the frames generated by the visual model. We update the components of this model in an alternate fashion, keeping the visual model constant during the training phase of the temporal model in order to allow it to identify the loss of valuable temporal information from the reconstructed frame. Similarly, we then keep the temporal model constant while the visual model is training to improve the results of its outputs when they are evaluated by the temporal discriminator.

This joint training allows our system to learn an embedding space that not only captures visual properties by predicting the next frame but also ensures that we retain enough temporal information to estimate a consistent timestamp for the output. The embedding space of the model is then used for clustering to form the respective classes and segment the videos accordingly. We illustrate our clustering and segmentation pipeline during testing in Figure 6.4. We discuss each step of the pipeline in the following sections.

6.2.1.1   Stage 1: Disjoint Training of the Visual and Temporal Models

Visual Embedding Model: The visual embedding part learns a low-dimensional embedding of the frame-based input vector. To this end we make use of an encoder-decoder structure to model a frame prediction framework, using as a prior the visual features of the frame at time t (frame f_t) and learning to predict the features at a future time $t + s$ (f_{t+s}).

As is shown in Figure 6.3a, our visual model (displayed in blue) consists of three down-sampling layers followed by three up-sampling layers with skip connections between the down-sampling and up-sampling layers to preserve fine-grained details from being discarded by the encoding process. As such, we are able to suppress visual noise and find visual cues that are more relevant to the respective task while ensuring that the encoded embedding x_t maintains enough relevant information in order to reconstruct a viable future frame in the video sequence. This linear U-Net is then used during testing to generate an abstract representation of the frame's features x_t (generated by the last down-sampling layer), which is then used as the embedding of the frame f_t during the later clustering process (see Figure 6.4).

This visual embedding model learns to generate accurate future frame predictions by minimizing the mean squared difference between the predicted and the real frames:

$$Loss_{vis} = \frac{1}{N - s} \sum_{t=1}^{N-s} (f_{t+s} - \hat{f}_{t+s})^2 \tag{6.1}$$

For ease of notation, we use N as the number of all frames of all videos V belonging to the same activity class A, and s as the prediction time step. That is, for $s = 5$ given an input frame captured at timestamp $t = 0$ ($f_{t=0}$) the network will return a predicted frame s frames into the future ($\hat{f}_{t=5}$). For our experiments, we used $s = 5$ as the prediction goal of our visual embedding model. However, we also study the impact of different step sizes and provide a comparative analysis of their impact on the action segmentation task.

Temporal Embedding Model: The temporal embedding model (displayed in red in Figure 6.3a) is designed to extract valuable temporal information out of a given frame. For this, we implemented a three-layer Multilayer Perceptron (MLP) with the learning goal of predicting the relative timestamp t of a given frame (where $t = frameIdx/N$). The MLP receives as input a video frame f_t captured at a timestamp t and provides a prediction for the relative timestamp of the frame $T(f_t)$. As such, to ensure the MLP has a working knowledge of the temporal structure of the data, it must minimize the difference between the timestamp prediction $T(f_n)$ it provides for a given f_t frame, and the actual timestamp t of such frame:

$$Loss_{temp} = \frac{1}{N} \sum_{t=1}^{N} (T(f_t) - t)^2 \tag{6.2}$$

6.2.1.2 Stage 2: Joint Visual-Temporal Training

In this stage we joined our visual and temporal models in a single framework, as such, our final architecture is composed of (1) a frame predicting U-Net that is able to retain enough visual and temporal features in its predictions, and (2) a timestamp predictor, trained to recognize any discrepancy between the temporal quality of the frame predicted by the U-Net $\hat{f}_{t+s}$ and the observed frame f_{t+s}.

Joint Visual-Temporal Embedding Model: The learning objective of the U-Net during stage 1 was to predict a frame that seemed visually plausible (Equation 6.1), this helps to ensure that the valuable visual cues in the frame embedding are retained. In the second stage of our training, we build on this objective to ensure that the temporal cues of the frame are also preserved. For this we use our temporal model: the timestamp predictor evaluates the quality of the temporal features encoded into our embedding by measuring the difference between the predicted timestamp of the U-Net's output $T(\hat{f}_{t+s})$ and the predicted timestamp of the observed frame ($T(f_{t+s})$, allowing us to measure the temporal reconstruction loss of our generated frames:

$$Loss_{trec} = \frac{1}{N-s} \sum_{t=1}^{N-s} (T(f_{t+s}) - T(\hat{f}_{t+s}))^2 \tag{6.3}$$

We incorporate this temporal reconstruction loss into the U-Net's learning objective:

$$Loss_{joint} = Loss_{vis} + Loss_{trec} \tag{6.4}$$

This joint loss (Equation 6.4) ensures that our embedding model maintains a good balance of visual and temporal cues into its embedding. Its learning objective is now minimizing both the visual and temporal reconstruction losses.

Temporal Discriminator: The goal of our temporal embedding model is now to serve as a discriminator for the reconstructed frames generated by our Joint Visual-Temporal Embedding Model. For this, our temporal discriminator must be able to identify the loss of the "temporal quality" between the frame predicted by the U-Net $\hat{f}_{t+s}$ and the ground-truth frame f_{t+s}. Given a frame f_t we consider that the U-Net's embedding x_t has led to the loss of temporal cues if the predicted timestamp $T(\hat{f}_{t+s})$ (estimated by the temporal embedding model) of the U-Net's output $\hat{f}_{t+s}$ is not similar to the timestamp of the ground-truth frame f_{t+s}. The temporal quality ($TQual$) of the U-Net embedding can then be measured as follows:

$$TQual = 1 - \left(\frac{1}{N-s} \sum_{t=1}^{N-s} T(\hat{f}_{t+s}) - (t+s))^2 \right) \tag{6.5}$$

We can then use the estimation of the temporal quality loss as a means for the MLP to discriminate between a real video frame f_{t+s} and a low-quality estimation $\hat{f}_{t+s}$. As a result, the timestamp predictor can learn to discriminate between the temporal differences of the U-Net's prediction and the ground-truth. The loss of the

MLP temporal discriminator is then evaluated as follows:

$$Loss_{MLP} = \frac{\frac{1}{N}\sum_{t=1}^{N}(T(f_t) - (t))^2}{TQual} \tag{6.6}$$

In the above formulation, $T(f_i)$ represents the output of the timestamp predictor given an input frame f_i.

6.2.1.3   Clustering and Decoding

For all further processing, we follow the protocol of [11, 85, 147]. Once we have the temporally enhanced embedding, we cluster the embedded features of all videos into k clusters which would later be mapped to the k sub-activities associated with the task type. For this, we employ a single Gaussian Mixture Model to model all the embedded features of a given task. To fit them into k clusters, we then use the logarithm of the probability of each frame belonging to each of the clusters. As such, for every video frame n, we have a corresponding k dimensional vector with each of the frame's scores for each of the clusters (see step 2 in Figure 6.4).

A naive approach for segmentation would be using the scores obtained by the clustering method in order to assign a sub-activity to each video frame. This, however, generates a non-homogeneous segmentation with few groups of continuous frames being assigned to the same sub-activity cluster. We use a modified version of the frame labeling method presented in [85], Viterbi decoding with length model as proposed by [84] to alleviate this effect. We evaluate the probability of each frame n belonging to a cluster c_x with respect to the probabilities of the neighboring frames and seek to maximize the probability of the sequence following a fixed cluster order. The cluster order is determined by the mean time-stamp of each cluster.

A fixed cluster ordering $c_1, c_2, ..., c_i, c_j, ..., c_k$, would constrain the possible clusters a frame f_t can be assigned to. Frame f_t could belong to either the same cluster c_i as it preceding sampled frame $f_{t-\gamma}$ -where γ is the frame sampling size (in the method introduced by [85] $\gamma = 1$) or to the next cluster c_j in the predetermined ordering.

6.2.2 Evaluation

6.2.2.1 Datasets

We evaluate our method using three datasets: Breakfast Actions dataset (BF) [82], INRIA YouTube Instructional Videos (YTI) [3], and 50 Salads (50S) [156]. We are using the reduced Fisher Vector features as proposed by [83] and used by [85, 130, 147] for the evaluation of the three datasets.

The Breakfast Actions (BF) dataset contains 70 hours of cooking activities of varying complexity. It contains 10 different cooking tasks (with about 170 videos per task), which can be further split into 48 sub-activities. The length of each video is highly dependent on the type of task, ranging from 30 seconds to a few minutes. The videos are recorded in different real-life environments with 52 people performing each of the 10 different actions. They have a fixed viewpoint through all activities for each person, which leads to a high intra-class and low inter-class variance.

The INRIA YouTube Instructional Videos (YTI) dataset contains five tasks of different instructional domains: "making coffee", "changing a car tire", "CPR", "jumping a car", and "potting a plant" (with about 30 videos per task), which can be divided into 47 sub-activities. As opposed to Breakfast Actions, the videos might have been edited and include shot boundaries, as well as different and changing viewpoints or zooming. The videos in this dataset are in average longer than the videos in Breakfast Actions, however, there is a significant presence of background frames whereas the Breakfast Action actions are densely labeled without intermediate background classes.

The 50 Salads (50S) dataset contains over 4.5 hours of video captures of different actors preparing 2 kinds of mixed salads (with 25 videos for each type of salad). While similar in nature to the data present in the Breakfast dataset, the videos in 50 Salads tend to be longer and contain more sub-activities. These videos have an average length of 10k frames and contain complex interactions between hands, utensils, and ingredients. In addition to this, different actions in this dataset often contain similar motions, differing only on the ingredient or utensil used and can be further classified into three stages: preparing the dressing, cutting and mixing ingredients, and serving. As such we have similar actions like "adding oil" or "adding vinegar" and "cutting tomato" or "cutting cucumber" which follow similar hand motions (but differ in the interacting objects) and can happen at interchangeable times.

6.2.2.2   Training

Following the two-stage training protocol described in Section 6.2.1, we first train the visual and temporal models independent from each other. The visual embedding model is trained for 160 (for the BF Dataset) and 140 epochs (for the YTI and 50S datasets) while the temporal embedding model is trained for 10, 20, and 30 epochs (for 50S, YTI, and BF respectively).

During our second training stage, we alternate the training of the visual and temporal models: we train the visual model for 40 consecutive epochs (in which the temporal model is frozen) and then train the temporal discriminator for 5 epochs (where the visual embedding model remains constant. We run these alternating periods for 60 (for the BF dataset) and 120 epochs (for the YTI and 50S datasets).

6.2.2.3   Ground-truth Mapping and Evaluation Metrics

To evaluate the segmentation returned by our unsupervised approach, we need to map the segmented clusters of sub-activities to the ground-truth sub-activities related to each specific task. For this, we use Hungarian matching to provide a one-to-one mapping that maximizes the similarity between the segmented clusters and the ground-truth sub-activities. We follow the protocol of [11, 85, 147] and compute the Hungarian matching over all the videos of one activity. Note that this is different from the Hungarian matching for every single video as used by [1], which optimizes the matching for every single video and usually leads to higher accuracy, but also allows clusters to change their label from one video to another.

We evaluate the accuracy of our method using two common evaluation metrics used in [1, 85, 147], 1) *Mean Over Frames (MoF)* to indicate the percentage of frames in the segmentation that were correctly labeled over all the frames of videos assigned to a given task for the Breakfast Actions and 50 Salads Datasets, and 2) *F1* score for the INRIA YouTube Instructional Videos (YTI) Dataset to compare with other state-of-the-art approaches.

6.2.2.4   Comparison to State-of-the-Art Methods

We compare the performance of our two-stage visual embedding pipeline to state-of-the-art approaches (including weakly-supervised and fully supervised setups) on three challenging action segmentation datasets. As can be seen in Figure 6.5 our approach is able to maintain the logical ordering of sub-activities and a good estimate of their start and duration in actions with different levels of complexity.

Breakfast Actions Dataset: Our method is able to outperform the state-of-the-art unsupervised learning methods applied to this dataset (see Table 6.4) and performs competitively against weakly-supervised approaches. It is important to note that our cluster-to-ground-truth mapping and evaluation is done globally, that is, we use all the predicted labels and ground-truth for all the videos of a given task to do the Hungarian matching, and then evaluate calculating the MoF using the count of all the true predictions on the whole dataset. On the other hand, approaches such as LSTM+AL [1] employ a per-video (local) cluster to ground-truth mapping,

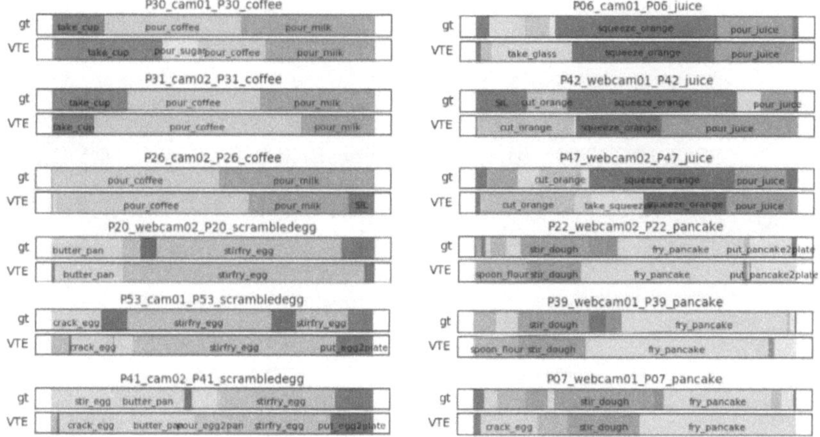

Figure 6.5. Segmentation examples from our approach on videos from the Breakfast Actions dataset. We omit some of the ground-truth labels in the visualizations to facilitate reading the labels of the more prominent segments.

which might account for the difference in the performance of the two approaches. For comparison purposes, when applying a video-based Hungarian matching our method obtains an MoF of 52.25%.

6.2.3 Results and Analysis

INRIA YouTube Instructional Videos (YTI): The YTI Dataset is particularly challenging as the majority of its frames consist of "background" information (see Figure 6.6). That is, most of the frames in this dataset do not provide any relevant visual cues related to the action we are segmenting. In line with other approaches, our evaluation of this dataset excludes any frames our method assigned to the label background to avoid any bias towards the over-segmentation of this sub-activity. Despite the reduced amount of valuable visual cues, we can gather from single frames in this data, our method outperforms other global-based Hungarian

TABLE 6.4

SEGMENTATION RESULTS ON THE BREAKFAST ACTIONS DATASET
COMPARED TO STATE-OF-THE-ART METHODS. (*DENOTES RESULTS
WITH VIDEO-BASED HUNGARIAN MATCHING)

Supervision	Approach	MoF
Full	TCFPN [32]	52.0%
	HTK [83]	56.3%
	GRU [137]	60.6%
	MS-TCNN++ [100]	67.6%
	Local SSTDA [18]	70.2%
	SSTDA[89]	**70.3%**
Weak	Fine2Coarse [136]	33.3%
	GRU [137]	36.7%
	TCFPN+ISBA [32]	38.4%
	NN-Viterbi [138]	43%
	D3TW [14]	45.7%
	CDFL [98]	**50.2%**
Unsupervised	GMM [147]	34.6%
	CTE-MLP [85]	41.8%
	(LSTM+AL [1])	(42.9%*)
	Ours	**48.08%**

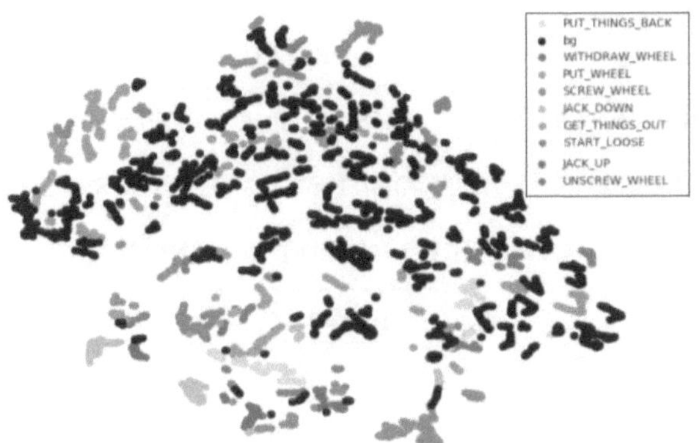

Figure 6.6. t-SNE[175] Visualization of the cluster distribution on a video from the YTI Dataset. Notice the heavy presence of background (bg) frames.

matching approaches (see Table 6.5). We observed a considerable improvement when using larger frame prediction steps (predicting further into the future) for this dataset, which probably helped our visual embedding model to discriminate the minutiae associated with irrelevant frames.

50 Salads (50S): While more similar in nature to the videos in Breakfast Actions, the order of the sub-activities in 50 Salads is not as important as with the previous datasets. Sub-activities belonging to the same category have similar temporal locations (e.g., "cutting tomato", and "cutting cucumber"), and often the order in which they are performed does not have a big impact on the performance of the task. Consequently, the benefits we see from following the two training stages (Table 6.6) are not as significant as for the other two datasets, just using the visual embedding from the first stage seemed to provide fewer distractions (brought on by the weak temporal ordering of the actions in this dataset) and resulted on a better performance.

TABLE 6.5

SEGMENTATION RESULTS ON THE YTI DATASET COMPARED TO UNSUPERVISED STATE-OF-THE-ART METHODS. (*DENOTES RESULTS WITH VIDEO-BASED HUNGARIAN MATCHING)

Approach	F1
GMM [147]	27.0%
CTE-MLP [85]	28.3%
(LSTM+AL [1])	(39.7%*)
Ours	**29.86%**

TABLE 6.6

MOF SEGMENTATION RESULTS ON THE 50 SALADS DATASET. (*DENOTES RESULTS WITH VIDEO-BASED HUNGARIAN MATCHING)

Supervision	Approach	Eval	Mid
Full	S-CNN [89]	68.0%	54.9%
	ST-CNN[89]	72.00%	58.06%
	ED-TCN [90]	73.4%	64.7%
	MS-TCNN++ [100]	–	83.7%
	Local SSTDA [18]	–	82.8%
	SSTDA[89]	–	**83.8%**
Unsupervised	CTE-MLP [85]	**35.5%**	**30.2%**
	(LSTM+AL [1])	(60.6%)*	-
	Ours	30.59%	24.19%

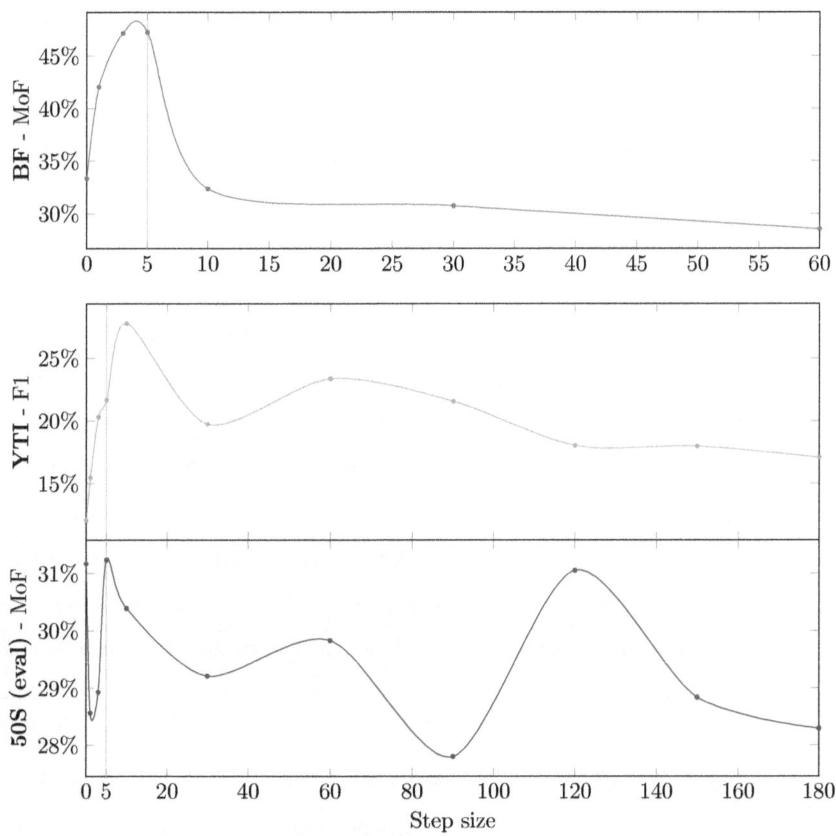

Figure 6.7. U-Net Embedding Performance when predicting longer step sizes (s). We do not test $s > 60$ for BF as some of the videos in the dataset have as few as 180 frames.

6.2.3.1   Future Frame Prediction Task

In order to assess the value of future frame prediction as a good learning task to encode both the visual and temporal cues of the data, we tested the performance of our visual embedding model using a frame prediction step sizes s of 0, 1, 3, and 5 (see Table 6.7). When $s = 0$ the network would learn to reconstruct the same input frame from the generated encoding (rather than generating a prediction of a future frame), with $s > 0$ the network would predict the next s_{th} frame in the video sequence following the input frame (we should be able to generate a given frame t from the encoding of a previously seen frame at time $t - s$).

We observed a distinctive improvement for both the BF and YTI Datasets as the step size increased, with the lowest performance obtained by just reconstructing the original input frame ($s = 0$). The strong correlation between temporal and visual cues in these datasets allowed us to extract a more significant encoding using our approach. However, it is important to note that, while our approach benefits from the future frame prediction task, when testing overly large step sizes (see Fig. 6.7) the performance degraded as the step size increased. Overly large step sizes could exacerbate the erosion of shorter sub-activities from the temporal segmentation as well as disrupt the learning of the logical ordering of the sub-activities in a given action.

This was not the case for the 50 Salads dataset, while overall the U-Net with the largest step size obtained the best performance, the further enforcement of the temporal encoding brought on by the second stage of our training pipeline proved to be detrimental as the temporal cues are not that significant in this type of data.

6.2.3.2   Visual vs. Temporal Embedding

The purpose of our embedding algorithm is to encode valuable visual and temporal information derived from our input features, to evaluate the impact of such embed-

TABLE 6.7

U-NET LEARNING TASK IMPACT ON THE BREAKFAST ACTIONS
DATASET. GIVEN AN INPUT FRAME CAPTURED AT A TIME t THE U-
NET WAS TRAINED TO ESTIMATE THE FRAME s STEPS AFTER THE
INPUT FRAME.

Step size	BF	YTI	50S (eval)
U-Net (s=0)	33.34%	12.16%	31.17%
U-Net (s=1)	42.06%	13.51%	28.56%
U-Net (s=3)	47.17%	21.32%	28.93%
U-Net (s=5)	**47.27%**	**25.72%**	**31.24%**

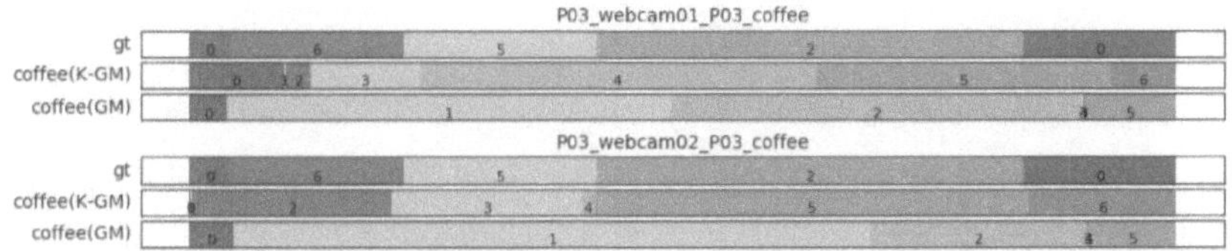

Figure 6.8. Segmentation illustration using different clustering methods on
the U-Net embedding. Notice that the GM clustering method has a tendency
of over-fitting to dominant classes.

ANALYSIS OF THE FEATURE EMBEDDING RESOURCES ON THE
BREAKFAST ACTIONS DATASET AT BOTH STAGES OF TRAINING.
STAGE 1 MODELS WERE TRAINED INDEPENDENTLY OF EACH OTHER.

Training	Embedding Source	MoF
–	Fisher Vector (no embedding)	28.13%
Stage 1	U-Net	47.27%
	MLP	40.91%
Stage 2	U-Net	**48.08%**
	MLP	39.04%

ding, we first analyze the performance of the segmentation pipeline when omitting the feature embedding step. For this, we skipped the embedding process and directly clustered the reduced Fisher Vectors following the process described in Sec. 6.2.1.3. We observed a significant decrease in the performance obtained for all three datasets, with a 13.5% F1 score for the INRIA Youtube Instructional Videos (YTI) Dataset, and an MoF score of 28.13% and 29.07% for the Breakfast Actions (BF) and 50 Salads (50S) Datasets respectively.

As described in Sec. 6.2.1, we train our visual (U-Net) and temporal (MLP) components in two stages to obtain a joint visual and temporal embedding (extracted from the visual model after the second stage). We now evaluate the impact of different types of embedding (visual, temporal, or joint) during both stages of our training. Table 6.8 shows the result using the embedding of the visual and temporal components, the U-Net and the MLP respectively. At stage 1 both components were trained in a disjoint fashion, following the losses described in Equations 6.1 (U-Net) and 6.3 (MLP). At stage 2 the loss of both models is updated so that they are trained jointly (alternating the training of the U-Net and MLP for 40 and 5 epochs respectively)

TABLE 6.9

COMPARISON OF THE EFFECT OF DIFFERENT LOSSES IN U-NET+MLP

EMBEDDING ON THE BREAKFAST ACTIONS DATASET.

U-Net+MLP Loss	MoF	IoU	Average
Visual	39.66%	9.97%	24.82%
Temporal	39.35%	11.94%	25.64%
Visual+Temporal	**42.66%**	**12.76%**	**27.71%**

following Equations 6.4 (U-Net) and 6.6 (MLP). We used the output of the U-Net's encoder as our embedding for the U-Net, and the output of the inner layer of the MLP as its embedding. Overall the visual embedding of the U-Net after the first stage already led to an increased performance of the whole system. Nevertheless, further enforcing a stronger temporal embedding through stage 2 provided the best performance for the Breakfast Actions Dataset.

6.2.4 Discussion

Various approaches have been proposed for learning visual representations without labels [1, 75, 85, 147]. Particularly, in the case of learning video representations in an unsupervised way, the temporal properties of the data have been commonly employed to understand the structure of the observed visual features and the rules they follow while they change over time. Following the idea that given a set of videos all capturing the same activity it should be possible to identify temporal segments with similar sub-actions across all videos, we have studied the suitability of a joint visual-temporal approach to learn to identify different action segments from unlabeled video data.

The results of our experiments have shown that a combination of predictive visual and temporal learning tasks encourages the encoding of valuable information for temporal segmentation that might have been otherwise disregarded by a solely temporal approach. The segmentations obtained with our method maintain the logical ordering of the analyzed tasks and are able to produce coherent segments that have led us to significant improvements over previous approaches that have exclusively used temporal embeddings.

We have observed that incorporating the visual cues that might be present in a frame into our joint approach has shown promise in the area of temporal segmentation, particularly for cases in which the sub-activities have either 1) *distinct visual* appearances (e.g., "cracking egg" vs. "buttering pan"), and/or 2) a *strong temporal coherence* (e.g., the "making coffee" sub-activities "pouring coffee" and "pouring milk" would have similar visual aspects, however, in most cases the actors in the videos pour the coffee (into the mug) first before adding milk to it), as was the case for datasets like Breakfast Actions and INRIA Youtube Instructions (YTI)). However, our approach struggles with data that has weak temporal coherence and low visual variance (like the data in 50 Salads). We suggest as a future research direction incorporating an evaluation of the temporal coherence of sub-activities, both to address this issue and to incorporate the segmentation of repetitive actions (which lead to inconsistent results in their temporal location).

CHAPTER 7

CONCLUSIONS AND FUTURE WORK

Through this **book**, we have explored the intersection between computa-tional photography and visual recognition. These two fields have made significant advances in recent years, however, they have been studied separately. Through this work, we argue that there is significant potential when integrating them in order to improve both fields.

Computational photography involves using different techniques to enhance or manipulate digital images, while visual recognition involves using machine learning algorithms to identify objects or patterns within images. As we have seen throughout this work, image enhancement can have positive and negative effects on neural network predictions. In some cases it can help improve the performance of machine learning architectures, in others, post-processing techniques can, either maliciously or accidentally, tamper with an otherwise correct response from a visual recognition system.

In general, image enhancement techniques aim to improve the quality of images by removing noise, sharpening edges, adjusting brightness, improving contrast, or other similar operations. These techniques can improve the visual appearance of images and make it easier for humans to perceive important details, while this can in some cases benefit a machine learning system, it can also introduce artifacts or distortions that may affect the performance of neural networks. For example, some enhancement techniques can increase the contrast between different regions of an image, making it easier for a neural network to distinguish between different objects or features.

On the other hand, some techniques can introduce high-frequency noise or amplify low-frequency noise, which can interfere with the neural network's ability to extract meaningful features from the image.

Overall, the effects of image enhancement on neural network predictions are complex and depend on many factors. It is important to carefully evaluate the impact of image enhancement techniques on specific neural network models and applications before incorporating them into a visual recognition pipeline. Our benchmark dataset, UG^2 has been a stepping stone toward this effort, this dataset and the evaluation metrics we introduced with it have been a major tool for evaluating the impact of image restoration and enhancement techniques on automatic visual recognition and have been used by researchers around the globe through our CVPR's UG^2 Challenge.

As part of this challenge, we addressed the importance of considering real-world scenarios when evaluating the performance of visual recognition algorithms. As observed in the UG^2 dataset chapter, the performance of state-of-the-art approaches on real-world noisy/corrupted data is far from that observed on good-quality data. Many of these degradations have been studied in the field of computational photography with the intention of improving the aesthetics of a given image, taking this into account we studied the effectiveness of different restoration and enhancement techniques in improving recognition performance.

We found that image restoration and enhancement techniques can significantly improve the performance of these systems. However, none of the evaluated enhancement approaches could be widely applied to all types of data. While some techniques drastically improved the results on highly corrupted media, these same approaches would have a negative impact on images with few to none-existing degradations. Real-world variability and complexity of scenes captured in unconstrained scenarios make it difficult for single-purpose image restoration and enhancement techniques to work well across all types of data. Taking this into account we proposed a joint enhancement and recognition framework, which would allow for the tuning of an enhancement technique with the purpose of improving the performance of a given visual recognition system, and highlighted the need for continued research in this area to address the many challenges associated with the processing of data with multiple sources of degradations.

Additionally, the impact of enhancement techniques on visual recognition systems can heavily depend on the specific task being performed and the amount of data required to perform such task. Tasks like object tracking or action recognition would often benefit from having access to multiple frames capturing the same scene, and as such could be more robust to certain types of degradations. These tasks can also benefit from more complex and temporal-based enhancements that could allow for a finer transformation of the frame's pixel-level and deep features.

To analyze the importance of temporal cues in conjunction with the visual information present in a scene, we studied the value of integrating both sources of information into a joint embedding. Following our proposed framework, we jointly trained this embedding system to tune these joint features to be better suited for an action segmentation task. We observed that this allowed us to generate a compact and discriminative representation for each frame in a video sequence, which aggregated into a video-level representation and allowed for a better representation of the actions being performed.

These changes in image characteristics can affect the deep features extracted by the neural network in several ways. For example, brightening an image can cause the network to focus more n the high-intensity regions of the image, while contrast enhancement can make the network more sensitive to edges and textures. However, these changes can also lead to the extraction of spurious or irrelevant features which can degrade the overall performance of the neural network.

Following the idea that the deep features extracted by a neural network can be affected by the changes in an image's visual characteristics, we studied the cases where these changes could also lead to the extraction of spurious or irrelevant features that would in turn degrade the overall performance of a network. For example, changes to an image's contrast could cause some features to become over-emphasized leading to a distortion of a network's perception of the scene. In this avenue, the impact of different image transformations has been an object of interest in the field of image forensics, with the area of manipulation detection paying special attention to local manipulations and how to find them.

While it is commonplace for manipulation detection techniques to look for minute traces in the pixel-level domain, there has been limited research on the benefits we could gain from incorporating some of these manipulation techniques to facilitate the identification of forgery. We argued these same techniques can be used to enhance the visibility of hidden or obscured details in images that could help identify evidence of tampering or forgery. Our proposed solution then is trained to enhance any present manipulations in an image and reduce the presence of low-confidence regions that could be ignored by a manipulation detection approach. Through multiple experiments, we have proved that the use of image enhancement techniques in a joint enhancement-recognition framework can improve manipulation detection for multiple detection approaches, regardless of the specific cues said approaches use to detect tampering.

Through this work, we studied the benefits we can gain from joining the fields of image enhancement and visual recognition. This is a promising research area that could integrate the strengths of these two fields to develop more effective and efficient techniques for capturing and processing visual data. The potential research directions can also include the use of visual recognition systems to improve the results of traditional enhancement approaches, such as using detection or segmentation techniques to improve or locally enhance different areas of an image based on their unique characteristics, or developing degradation specific techniques to globally address issues across intersecting tasks such as object classification, detection, or segmentation. Visual recognition algorithms can also be used to improve image enhancement approaches by providing feedback on the effectiveness of different enhancement techniques in improving recognition performance. Specifically, visual recognition algorithms can be used to evaluate the impact of different image enhancement techniques on recognition accuracy and to identify which techniques are most effective at improving recognition.

Overall the integration of computational photography and visual recognition algorithms in the context of object detection and recognition is an exciting area of research with many promising directions for future work.